THE FIRST CAMPAIGN

ADVANCE OF GENERAL ROSECRANS'S DIVISION THROUGH THE FORESTS, TO ATTACK THE CONFEDERATES AT RICH MOUNTAIN.

THE
FIRST CAMPAIGN

A Guide to Civil War in the
Mountains of West Virginia, 1861

THREE ONE-DAY DRIVING TOURS

HUNTER LESSER

*To Jim Skidmore,
desendant of Patroits in the
31st Virginia Infantry*

QUARRIER PRESS
Charleston, West Virginia

Quarrier Press
Charleston, WV

First Edition

10 9 8 7 6 5 4 3 2 1

Printed in China

Library of Congress Control Number:
2011931949
ISBN-13: 978-1-891852-75-6
ISBN-10: 1-891852-75-2

All modern photos by Hunter Lesser unless
otherwise noted.

Book design: Mark S. Phillips

Distributed by:

West Virginia Book Company
1125 Central Avenue
Charleston, WV 25302
www.wvbookco.com

Acknowledgements

Alderson-Broaddus College
Beverly Heritage Center
Terry Hackney
Hargrett Rare Book & Manuscript Library,
    University of Georgia
The Huntington Art Collections
Indiana Historical Society
Library of Congress
Library of Virginia
Mary Baldwin College Archives
Michael Ledden
Walt Lesser
Jessie Beard Powell
Randolph County (WV) Historical Society
Rich Mountain Battlefield Foundation
Ed Riley and Donna Depue
Robert and Anita Schwartz
William A. Turner
USDA Forest Service, Monongahela National
    Forest
West Virginia and Regional History Collection,
    West Virginia University Libraries
West Virginia State Archives

Support the battlefields:

www.richmountain.org;
www.battleoflaurelhill.org;
www.corricksford.org

Table of Contents

Camp Allegheny

"*The whole region is wild and grand, and if any one of the men who in his golden youth soldiered through its sleepy valleys and over its gracious mountains will revisit it in the hazy season when it is all aflame with the autumn foliage I promise him sentiments that he will willingly entertain and emotions that he will care to feel. Among them will be, I fear, a haunting envy of those of his war comrades whose fall and burial in that enchanted land he once bewailed.*"

—Ambrose Bierce, 1903

PROLOGUE

The secret is out. One of the great dramas of American history has been rediscovered. The **First Campaign of the Civil War**, set in the picturesque mountains of West Virginia (known as "Western" Virginia in 1861), has long been an overlooked chapter of our past. This guidebook, made up of three one-day driving tours, will help you explore that turbulent time.

Renewed interest in this story has been sparked by my book, *Rebels at the Gate: Lee and McClellan on the Front Line of a Nation Divided* (Sourcebooks, 2004), and by the work of many historians, interpreters, Civil War re-enactors, and preservationists.

Whether you plan to tour these "enchanted" mountains, or simply wish to enjoy a compelling tale, please read on!

HELPFUL TIPS

These tours originate in Elkins, West Virginia, gateway to the Allegheny Highlands. Directions (color coded) are listed for each day of the tour; please read them before departing to avoid problems. The mileage between destinations is approximate; odometers may vary, but you should be close!

Total Mileage by Day
Day 1 (Option 1) 100 miles
        (Option 2) 97 miles
Day 2 102 miles
Day 3 (from Elkins) 104 miles

When following the **Directions**, all mileage, unless otherwise indicated, is from the starting point of each day.

Please:
1. Carry a compass; cardinal directions are given to guide you by car and on foot.
2. Dress appropriately. Temperatures can be much cooler in the mountains! Snow and ice may be present.
3. Drive defensively. You will be traveling on curvy mountain roads.
4. Respect private property. Do not trespass unless public access is indicated!
5. Do not walk on or disturb earthworks and cultural features. State and Federal law prohibits the digging and removal of artifacts!
6. Do have a wonderful trip!

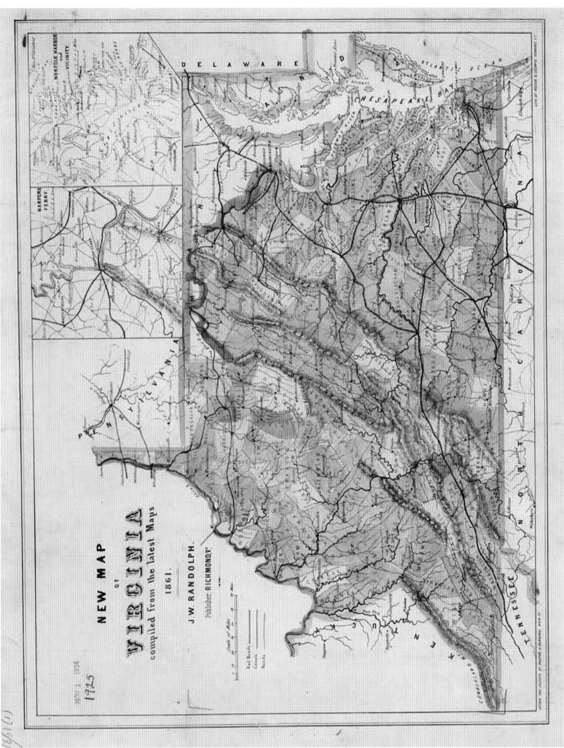

WEST VIRGINIA THEN & NOW
1861 Roads and Railroads
Contemporary Interstate Highways

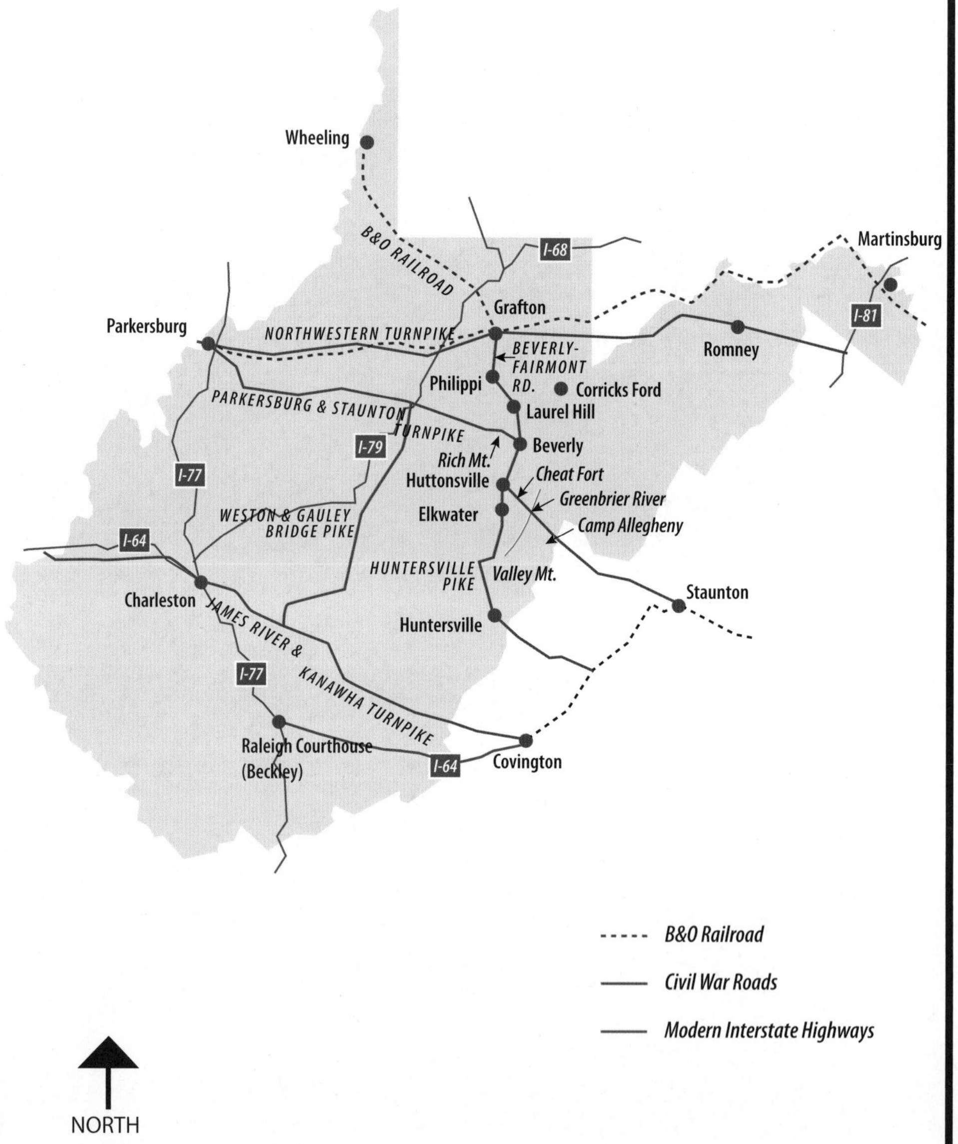

Wheeling

B&O RAILROAD

I-68

Martinsburg

Parkersburg

NORTHWESTERN TURNPIKE

Grafton

BEVERLY-FAIRMONT RD.

Romney

I-81

Philippi

Corricks Ford

Laurel Hill

PARKERSBURG & STAUNTON

TURNPIKE

I-79

Beverly

Rich Mt.

Huttonsville

Cheat Fort

I-77

Greenbrier River

WESTON & GAULEY
BRIDGE PIKE

Elkwater

Camp Allegheny

I-64

HUNTERSVILLE
PIKE

Valley Mt.

Staunton

Charleston

JAMES RIVER &

Huntersville

I-77

KANAWHA TURNPIKE

Raleigh Courthouse
(Beckley)

I-64

Covington

- - - - *B&O Railroad*

——— *Civil War Roads*

——— *Modern Interstate Highways*

NORTH

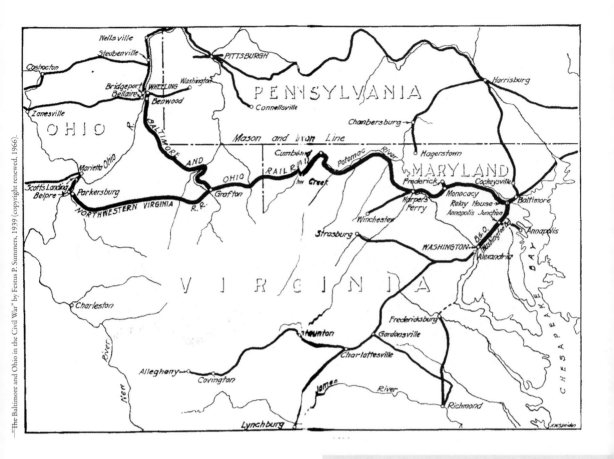

"The Baltimore and Ohio in the Civil War" by Festus P. Summers, 1939 (copyright renewed, 1966).

Virginia begins the War.

The Baltimore and Ohio Railroad— Lincoln's Lifeline

The B & O Railroad is one of the oldest in America. Begun in 1828, it reached Wheeling, Virginia by 1853. A branch line connected Grafton to the Ohio River at Parkersburg in 1857. During the Civil War, the B & O was vital to the Northern states. Thousands of Union troops and supplies passed over this railroad. It was a target of Confederate raids throughout the war.

INTRODUCTION:
Virginia Divided

On April 17, 1861, Virginia left the Union and promptly aligned with the Confederate States of America. The move sparked outrage in many of Virginia's western counties. Residents of "Western" Virginia had long sought a government of their own.

Virginia was divided by geography, culture and economics. The rugged Allegheny Mountains were a barrier to commerce between east and west. Westerners had stronger ties to the north, and less use for slavery. They resented a state government geared to the slaveholding east.

In 1861, the great issues that split North and South divided Virginians as well. So it was fitting that the **First Campaign** of America's Civil War would

The rugged mountains of Western Virginia, 1861.

be fought in the mountains of Western Virginia. These mountains became a proving ground for armies and leaders who shaped American history. George McClellan, Robert E. Lee and a host of others began the war here.

The First Campaign also led to the creation of a new state—West Virginia.

*"I hope to secure Western Virginia to the Union"—**George McClellan to Abraham Lincoln***

Tray Run Viaduct on the Baltimore and Ohio Railroad, near Rowlesburg. Confederate General Robert E. Lee thought the destruction of this vital crossing would be "worth to us an army."

The First Campaign

By May 1861, Confederate soldiers gathered near the Baltimore and Ohio Railroad junction at Grafton, Taylor County, Western Virginia. Meanwhile, Union volunteers mustered at Cincinnati, headquarters of the Army of the Ohio. Major General George B. McClellan, commanding Ohio forces, received letters from Western Virginia Unionists pleading for aid.

McClellan urged restraint, but on the night of May 25, Confederates burned bridges on the Baltimore and Ohio Railroad west of Grafton. The B & O was a vital Northern link—the main trunk line from Baltimore and Washington D.C. to the Midwest. This act triggered the **First Campaign**.

On May 27, 1861, General McClellan ordered Federal troops to cross the Ohio River on steamboats and invade Western Virginia. McClellan sought to protect the vital railroad and to rescue loyal Unionists.

Leading his invasion was the First Virginia Infantry, a regiment of *United States* volunteers, mustered on Virginia soil. Here the war was truly brother against brother.

*"Soldiers!—You are ordered to cross the frontier and enter upon the soil of Virginia. Your mission is to restore peace and confidence, to protect the majesty of the law, and to rescue our brethren from the grasp of armed traitors." —**Major General George McClellan***

Union Major General George McClellan. The First Campaign propelled him to stardom.

The Wheeling Custom House, seat of the Restored Government of Virginia.

George B. McClellan: The "Young Napoleon"

Born in 1826 of a distinguished Philadelphia family, George Brinton McClellan was star of the West Point class of 1846, a Mexican War hero, and a railroad president. The thirty-four year old McClellan took command of Ohio forces at the outbreak of the Civil War. President Lincoln soon appointed him a major general in United States service—outranked only by General-in-Chief Winfield Scott.

In a matter of weeks, McClellan built an army of nearly 20,000 Federal soldiers, invaded Western Virginia, and won the first Union victories of the war. Suddenly he was a "Young Napoleon," the North's first battlefield hero.

The "Restored Government" of Virginia

On May 13, 1861, a convention of Unionists gathered in the Virginia panhandle at Wheeling. By June 20, those delegates created the Restored Government of Virginia, a Union government to oppose the Confederates in Richmond. This daring deed sparked a movement toward statehood. On June 20, 1863, West Virginia became the nation's 35th state. She is truly a "child of the storm," forged amid the drama of the First Campaign.

Some Civil War "Firsts:"
Western Virginia, 1861

- First U.S. enlisted man killed by a Confederate soldier: T. Bailey Brown, May 22, 1861
- First regiment mustered on Southern soil for defense of the Union: First (U.S.) Virginia Infantry, May 23, 1861.
- First campaign of the Civil War: Federal troops under General George McClellan invade Virginia, May 27, 1861.
- First land battle of the Civil War: Philippi, June 3, 1861.
- First amputation of the Civil War: James E. Hanger, C.S.A., June 3, 1861.
- First use of the telegraph by an American army in the field: June 1861.
- First Union government restored in a Confederate state: Wheeling, June 20, 1861.
- First general officer killed in the Civil War: Robert S. Garnett, C.S.A., July 13, 1861.
- First time General Robert E. Lee leads troops into battle, September 1861.

William A. Turner

A typical volunteer soldier of 1861.

Citizens Go to War

At the outbreak of war in 1861, volunteers rushed to join the Union and Confederate armies. They were true "citizen-soldiers," from every walk of life. Some enlisted for three-months' service, as few on either side imagined a four-year conflict. The volunteers donned uniforms of every style and color. Union troops often wore gray, and Confederates dressed in blue—sometimes with tragic results!

Many carried antiquated firearms. Some Confederates fought with flintlock muskets, a weapon dating from the American Revolution. Others sported huge side knives. "Our large bloody-looking knives were the only things possessing much similarity," wrote one Confederate, "and a failure to have one of these pieces of war cutlery dangling at your side was almost a certain sign of weakness at the knees."

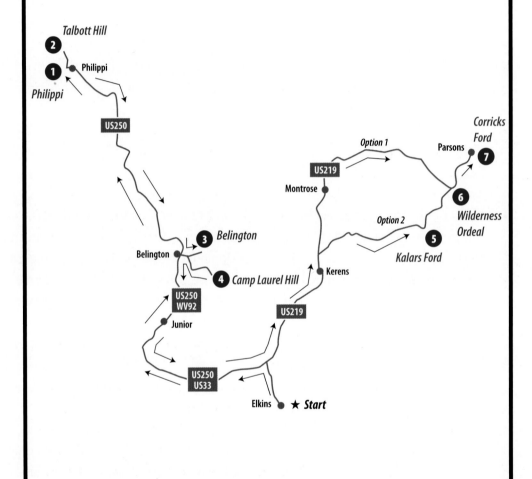

DAY 1
"Philippi Races" • Belington • Camp Laurel Hill ᵃ Corricks Ford

Day 1 Mileage
Option 1: 100 miles
Option 2: 97 miles

NORTH

The "Philippi Races," first land battle of the Civil War.

DAY 1

THE "PHILIPPI RACES," ACTION AT BELINGTON AND CAMP LAUREL HILL, BATTLE OF CORRICKS FORD

Preview: Day 1 opens the First Campaign, as volunteers rush to arms. On May 27, 1861, Federal troops under General George McClellan invade Western Virginia, crossing the Ohio River to secure the vital B & O Railroad and protect loyal Unionists. Badly outnumbered Confederates withdraw from the railroad at Grafton. On June 3, those Southerners are surprised at the "Philippi Races," first land battle of the Civil War.

The Confederates retreat south to Huttonsville, adding reinforcements and a new commander, General Robert S. Garnett. To slow the Federal juggernaut, General Garnett seizes two key mountain passes, gambling that McClellan will strike him at Laurel Hill. Instead, the armies skirmish there while McClellan defeats a Confederate force at Rich Mountain, 20 miles south, cutting off Garnett's line of retreat. The Confederates are trapped!

Garnett leads his Confederates northeast on rough mountain roads in a daring bid to escape. A Federal strike force pursues him. On July 13, a running fight breaks out along Shavers Fork of Cheat River. To save his army, General Garnett posts Confederate defenders at Corricks Ford. Here Garnett dies on the riverbank—the first Civil War general to fall. Humor and pathos, surprising Civil War "firsts," scenic rivers and mountain splendor await you on **DAY 1**.

> **Directions:** Elkins, the seat of Randolph County and a gateway to outdoor recreation, will be the hub for your tour. **Days 1 and 2 begin at the "Iron Horse," a landmark equestrian statue of founder Henry G. Davis, located on Randolph Avenue directly beside the Davis Memorial Presbyterian Church (US Routes 219/250/33) in Elkins. Set your odometer here**.
>
> From the "Iron Horse" statue, proceed north, go through one light, and get in the right lane. Turn right at the second light (0.1 mile) and continue on US 33 West (marked "To I-79") and turn left (2.0 miles) to enter the divided highway, US 33. Continue to the intersection of US 250/WV 92 (8.7 miles), turn right and continue north on the two-lane road, following US 250 through the towns of Junior and Belington to Philippi.
>
> At Philippi, (25.5 miles), get in the left lane and turn left on US 250 North (Truck). You will soon cross a bridge over the Tygart Valley River, then turn right on US 250/119 North (26.4 miles). Immediately turn right again, into Blue and Gray Park. Welcome to Philippi, site of the first land battle of the Civil War!

Stop 1—Philippi: Hotbed of Secession

In 1861, Philippi (seat of Barbour County) was considered by many to be the strongest Confederate town in Western Virginia. A "Palmetto Flag" flew over the courthouse early that year in support of South Carolina's secession from the Union. View the historic Civil War flags of Philippi at Blue and Gray Park.

The Philippi Covered Bridge

The large covered bridge spanning the Tygart Valley River was erected in 1852. It replaced a tedious ferry crossing on the Beverly-Fairmont Turnpike, leading north to the B & O Railroad at Grafton. More than three hundred feet in length and framed by massive rough-hewn logs, the "Monarch of the River" was built by a Beverly carpenter named Lemuel Chenoweth. This bridge symbolized the divided sentiments in Western Virginia: Chenoweth, a Secessionist, erected the span, while Emmet O'Brien, a Unionist, laid the stone foundation piers!

Amateur Armies

On May 27, 1861, Major General George McClellan's Union Army of the Ohio—almost 20,000 strong—crossed the Ohio River and invaded Western Virginia. McClellan's army was made up mostly of Ohio and Indiana volunteers, with some loyal Union regiments from Western Virginia. In response, a small

Philippi's covered bridge has survived floods, ice jams, war, and fire. The restored bridge remains the only two-lane covered span in America serving a Federal highway.

force of Confederates near the B & O Railroad at Grafton (fifteen miles north) retreated to Philippi.

Colonel George Porterfield, a VMI graduate and Mexican War veteran, commanded these Confederates. Porterfield's soldiers were poorly armed and equipped. They melted lead pipe to make bullets, and slept in the courthouse and other dwellings due to a shortage of tents. Colonel Porterfield's recruiting efforts were a disappointment. By June 1, no more than 775 Confederate volunteers had joined him at Philippi.

*"Philippi was a pandemonium. No order, our drill foolishness. The whole thing a holiday, full of disorder, uproar, speeches and intense excitement."—**John R. Phillips, 31st Virginia Infantry***

Philippi: First Land Battle of the Civil War

On the evening of June 1, trains bearing Federal troops rolled into Grafton, less than a day's march north. Leading them was General Thomas A. Morris of Indiana, a West Point graduate. At Grafton, General Morris found Colonel Benjamin F. Kelley preparing an attack.

Colonel Kelley, a former militia officer, commanded the First (U.S.) Virginia Infantry, a regiment made up in part of loyal Virginians who had mustered at Wheeling. Kelley was anxious to lead them against the "traitorous" Virginians at Philippi.

General Morris added his newly arrived regiments to Kelley's force, in a two-pronged movement to entrap the Confederates at Philippi. Colonel Kelley's column, 1,500 strong, left the B & O Railroad at Thornton (six miles east of Grafton) on the morning of June 2, and began a twenty-two mile march to Philippi. Their route lay along back roads east of the Tygart Valley River.

Colonel Ebenezer Dumont of the Seventh Indiana Infantry, a sallow Mexican War veteran, led the second column. On the evening of June 2, Dumont boarded the B & O at Grafton with 1,500 soldiers

Colonel George A. Porterfield, C.S.A. One Confederate called him "a polished Virginia gentleman, but as ignorant of war as a mule is of the Ten Commandments."

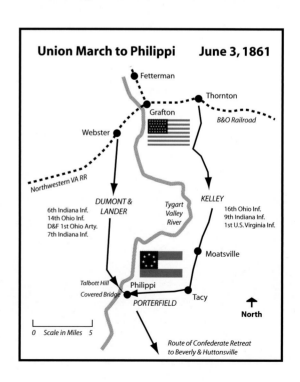

Union March to Philippi   **June 3, 1861**

and rode six miles west to Webster. Under cover of darkness, he struck the Beverly-Fairmont Turnpike for Philippi, twelve miles south. Dumont's orders were to engage the Rebels at dawn on June 3—just as Colonel Kelley cut off their retreat on the turnpike south of town.

Guiding Dumont's infantry was Frederick Lander, a flamboyant western explorer and volunteer *aide-de-camp*. Lander held the honorary title of "Colonel" as he led Dumont's men through the darkness toward Philippi.

It was their first march of the war. Drizzling rain soon became a downpour, turning the narrow country roads into quagmires. The night was pitch black. Federal soldiers slogged over the hills, tracing their progress by the frequent flashes of lightning.

"Talking, except in an undertone, was prohibited. The hours passed wearily away. Several halts for rest were made, and each time Col. Dumont [spoke] to each company of certain victory." **—G. W. H. Kemper, 7ᵗʰ Indiana Infantry**

Spies in Hoop Skirts

The Confederates had been alerted. Spies Abbie Kerr and Mollie McCleod rode into Philippi to warn Colonel Porterfield that a Federal assault was imminent. Mrs. George Whitescarver, wife of a Confederate soldier there, also raised the alarm. Fearful citizens packed up and fled the town.

But Colonel Porterfield chose not to retreat in the midst of the storm. Instead, he posted pickets on the roads leading to Philippi and went to bed. By midnight, his soggy sentinels returned to quarters. "Hell," exclaimed one Confederate, "any army marching tonight must be made up of a set of damned fools!"

Directions: Return to your vehicle. Leaving Blue and Gray Park, turn right on US 250/119 and proceed north, past the covered bridge and a stop light to the entrance of Alderson Broaddus College (27.1 miles). Turn left, and then promptly take another left into the main campus. Park in the Visitor Parking area (27.4 miles), adjacent to Burbick Hall (formerly the "Old Main.")

From this point, follow the sidewalk to the left and rear of Burbick Hall, where an interpretive sign and stone patio overlook the town of Philippi at the crest of Talbott Hill.

Stop 2—Talbott Hill

By 4:15 on the morning of June 3, 1861, Federal troops closed in on Philippi. Colonel Ebenezer Dumont's column, falling behind schedule, threw off their knapsacks and covered the last two miles of the march in "double quick" time.

"Colonel" Frederick Lander, the volun-

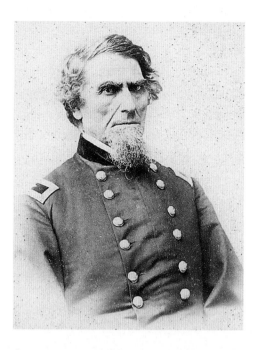

Benjamin F. Kelley, colonel of the First Virginia Infantry—a regiment fighting for the Union. He led the attack at Philippi.

A Woman Fires the First Shot

As Colonel Dumont's infantry filed by a house on the way to Talbott Hill, their clatter awoke Matilda Humphreys. Spying the shiny brass "U.S." buckles on their belts, Mrs. Humphreys put her twelve-year-old son Oliver on a horse, with instructions to warn the Confederates (an older son was in the army at Philippi). When Dumont's men pulled the youth from his saddle, Matilda Humphreys drew a pistol and fired.

Her wayward shot may have triggered the first land battle of the Civil War. As it echoed against Talbott Hill, Colonel Lander snapped erect and ordered his gunners to "Fire!" A young German immigrant named Lewis (Ludwig) Fahrion discharged the first bronze cannon, landing a ball squarely among the white tents five hundred yards below. As the artillery roared, Colonel Dumont's infantry stormed down a winding road and through the covered bridge.

The "Philippi Races" (June 3, 1861)

The Confederates in Philippi were taken completely by surprise. Jolted from bed by the novel sound of artillery, they spilled into the streets. Men and horses bolted down Main Street, almost trampling each other in their effort to get away. Some fled without boots, hats, coats, or pants! It was a "genuine shirt-tail retreat."

teer aide, led them to these heights, known as Talbott Hill. Two bronze 6-pounder guns of the First Ohio Light Artillery were wheeled to the front and unlimbered on the crest where you stand. The rain had ceased, and a dense fog obscured the village below. The gunners took their posts, awaiting a signal shot from Colonel Kelley on the other side of the river.

Soon the fog began to lift. The town of Philippi appeared below—the meandering river, the sturdy covered bridge, and a beckoning row of white tents. But Colonel Kelley's column was nowhere to be seen. The hour to attack (4 a.m.) was past. The Rebels were beginning to stir.

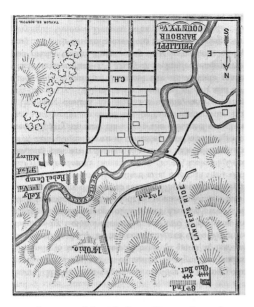

Early map of the Philippi Races, June 3, 1861

"*Out they swarmed, like bees from a molested hive. This way and that the chivalry flew, yet scarcely knew which way to run.*" —**Member of the Cleveland Light Artillery**

As the Confederates skedaddled out of Philippi, Colonel Kelley's Federals poured in. Misled by their guide, Kelley's column was fifteen minutes late, and on the wrong end of town, just east of the covered bridge. Soldiers of the First (U.S.) Virginia Infantry failed to catch their traitorous brothers. Putting spurs to his horse, Colonel Kelley charged alone through Philippi, firing a pistol at the fleeing Rebels until a Confederate quartermaster shot him from the saddle.

Colonel Lander plunged down the steep face of Talbott Hill on horseback to reach the scene. As Kelley's men prepared to bayonet his assailant, the gallant Lander intervened. Lander ordered them to pursue the fleeing Rebels, but they were worn out from the all-night march. With no cavalry at hand, the Federals were content to declare victory. The Confederates escaped to fight another day.

"*I must confess that I never saw a flight... executed with more despatch. They're not much for fight, but the devil on a run!*" —**Colonel Ebenezer Dumont, 7th Indiana Infantry**

"*[T]here was nothing left for us to do but to get out of town quickly, we all would have been captured that day were it not for the fact that the flanking columns lost their way and the attack on our flank and rear was not made.*" —**John H. Cammack, 31st Virginia Infantry**

So ended the "Philippi Races." Newspapers called it the "first land battle" of the war. Only a few shots were fired, and no one was killed. The casualties amounted to less than a dozen. Even the "mortally wounded" Colonel Kelley recovered and got a general's star for his heroics!

The First Amputation

The most noted Confederate casualty at Philippi was James Hanger, an eighteen-year-old member of the Churchville Cavalry. The third Federal cannonball fired into the town crashed through a barn and shattered Hanger's left leg. He was found bleeding in the hayloft, and was carried to

www.hanger.com

James E. Hanger, the Civil War's first amputee. His patented artificial leg spawned an industry.

the Philippi Methodist Episcopal Church where Dr. James Robison of the 16th Ohio Infantry cut off the mangled limb—without benefit of anesthesia. It was the first amputation of the war.

While recuperating, Hanger devised an artificial limb from barrel staves. His contraption worked so well that the "Hanger Limb" was patented. J. E. Hanger Inc. (now Hanger Orthopedic Group) became the largest manufacturer of artificial limbs in America and remains so to this day.

Aftermath

"I feel all right, and have come to the conclusion that I can stand almost anything, *and go through any privation. I have seen the elephant!"—G. W. H. Kemper, 7th Indiana Infantry*

The Philippi Races gave raw Union volunteers a first taste of success, and their presence in Western Virginia bolstered Union sentiment. Loyal Unionists soon gathered in a second convention at Wheeling. There, on June 20, 1861, the "Restored Government of Virginia" was born. A Union government was now in place to contest the Secessionists in Richmond!

Meanwhile, the Confederates withdrew to the Randolph County hamlet of Huttonsville. A court of inquiry later found Colonel Porterfield guilty of negligence. But General Robert E. Lee, commanding Virginia forces, ruled out a court-martial. Lee hoped "that the sad effects produced by the want of forethought and vigilance in this case" would be a lesson to be remembered.

☞ **Fun Fact:** Federal troops found Confederate Colonel William Willey in a sickbed at Philippi. "Bridge-burner" Willey had led the railroad vandalism that triggered the First Campaign. When discovered, he masqueraded as his half-brother Waitman Willey, a leading Unionist. The ploy worked until a Confederate commission and incriminating letters were uncovered!

FRANK LESLIE'S ILLUSTRATED NEWSPAPER.

DARING RIDE OF COLONEL LANDER AT THE BATTLE OF PHILIPPI.—FROM A SKETCH BY OUR SPECIAL ARTIST ACCOMPANYING MAJOR-GENERAL M'CLELLAN'S COMMAND.—SEE PAGE 102.

Frederick Lander's daring ride at Philippi.

Directions: Return to your vehicle. Go back to US 250, turn right (south) and return to the Philippi covered bridge. Turn left and drive through the bridge, noting the original arched beams. Upon exiting the bridge (28.3 miles), look left and note the 1909 railroad depot, home of the Philippi Museum. Here you will find images, artifacts, and publications on the bridge, the Battle of Philippi, and other local history. Don't forget to check out the Philippi Mummies, a real curiosity! (*www.philippi.org*)

After your tour of Philippi, leave the Philippi Museum parking lot and continue south on US 250 to the intersection of WV 92 at Belington (39.9 miles). Continue through this intersection on US 250, through a stoplight and turn left on Watkins Street (40.3 miles). Note the Laurel Hill Battlefield sign. Bear right at S. Beverly Street and immediately turn left on Serpell Avenue (40.4 miles). Continue to the Right Cemetery (40.8 miles) on your right. Park here, and walk the short, tree-covered lane through the cemetery to a large interpretive sign on the edge of the woods.

Stop 3—Belington: Battle at Laurel Hill (July 7-11, 1861)

Robert S. Garnett: A Dreary-Hearted General

Following the "Philippi Races," Confederate troops fell back to Huttonsville in the upper Tygart Valley. On June 14, General Robert S. Garnett arrived to take command. Garnett was a West Point man, forty-one years old, reserved and ramrod-straight—one of the most talented soldiers in the Confederacy. Robert E. Lee had sent him from Richmond to halt the Federal advance into Western Virginia.

Garnett was badly outnumbered. The Confederates under his command would not exceed 5,300, yet he was ordered to stop McClellan's army of nearly 20,000! Garnett hoped to rupture the B & O Railroad, but his outlook was grim.

"They have not given me an adequate force. I can do nothing. They have sent me to my death."—**General Robert S. Garnett, CSA**

The Beverly-Fairmont Turnpike

Completed in 1850, this road was a north-south link between the Staunton-Parkersburg Turnpike and the Baltimore and Ohio Railroad. From Beverly, former seat of Randolph County, the road wound north through Belington and Philippi to the B & O main stem at Fairmont.

Two mountain passes were the keys to Garnett's defense; one on the Staunton-Parkersburg Turnpike at Rich Mountain, the other on the Beverly-Fairmont Turnpike at Laurel Hill.

On June 15, General Garnett marched his little army north from Huttonsville. He left 1,300 Confederates in a strong pass at the western base of Rich Mountain. The next day, he occupied the pass over Laurel Hill, one mile east of this point. Garnett believed the ground at Laurel Hill was more vulnerable to attack. There, with the bulk of his army, he established headquarters.

"We are anxious to meet the foe, for we have them to whip, and the sooner we do it, the sooner we will be able to return to the dear loved ones at home." **—John B. Pendleton, 23rd Virginia Infantry**

McClellan's "Cerro Gordo"

On June 21, 1861, Union Major General George McClellan personally entered Western Virginia. His arrival sparked great fanfare. McClellan hoped to pull a "Cerro Gordo" on Garnett, duplicating a brilliant flanking movement by General Winfield Scott in the Mexican War. By the first week of July, McClellan led three Federal brigades toward Rich Mountain, intending to sweep around the Confederate left flank and seize Beverly, thereby cutting off Garnett's retreat.

Confederate General Robert S. Garnett — the assignment in Western Virginia seemed to him a death warrant.

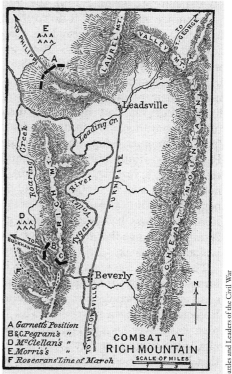

A *Garnett's Position*
B&C *Pegram's "*
D *McClellan's "*
E *Morris's "*
F *Rosecrans'Line of March*

COMBAT AT
RICH MOUNTAIN
SCALE OF MILES

Confederate positions at Laurel Hill (A) and Rich Mountain (B). These turnpike passes were General Garnett's "gates to the northwestern country."

27

Action at Laurel Hill (above). Federal troops in the foreground skirmish with a detachment of Confederates.

The Huntington Art Collections

*"Soldiers! I have heard that there was danger here. I have come to place myself at your head and to share it with you. I fear now but one thing—that you will not find foemen worthy of your steel."—**Major General George B. McClellan***

Ambrose Bierce—this Indiana volunteer, later a famous writer and critic, rescued a wounded comrade during the clash at Laurel Hill.

Baptism of Fire

On July 7, General Thomas Morris led a brigade of nearly 4,000 Federal troops from Philippi to the village of Belington. His orders were to *amuse* the Confederates here, to occupy them while General McClellan flanked the enemy at Rich Mountain, twenty miles south. Over the next four days, the armies skirmished along the Beverly-Fairmont Turnpike (County Route 15), winding between two cone-shaped hills in front of you. This area was the scene of spirited fighting. Here soldiers of the Blue and Gray received their baptism of fire.

*"A few dozen of us, who had been swapping shots with the enemies' skirmishers, grew tired of the resultless battle, and by a common impulse—and I think without orders or officers—ran forward into the woods and attacked the Confederate works. We did well enough considering the hopeless folly of the movement, but we came out of the woods faster than we went in—a good deal."—**Ambrose Bierce, 9th Indiana Infantry***

*"The whistling of musket balls and the peculiar note of the Minnie projectiles as they rush madly past on their errand of death is a frightful sound to the recruit who for the first time hears it."—**Colonel William B. Taliaferro, 23rd Virginia Infantry***

After two days of dueling, the Federals under General Morris captured those heights. One was named "Girard Hill," for the first Union soldier to fall upon its crest. His body was found in a makeshift grave, with the hand still exposed!

Directions: Return to your vehicle. Backtrack on Serpell Ave. for .2 miles (41.0 miles) to Judson Street. Turn left, and proceed .2 miles to the Beverly-Fairmont Turnpike (County Route 15) (41.2 miles). Turn left, and carefully continue southeast along this winding one-lane country road. As you pass the contested hills, note a small valley to your right, the scene of fighting during the engagement here. Confederate Camp Laurel Hill can be seen in the distance ahead, a cleared ridge trending perpendicular to the turnpike.

You will follow the Beverly-Fairmont Turnpike (County Route 15), past State historical markers for "Camp Laurel Hill." Turn right into the parking area at the City of Belington reservoir (43.0 miles). Park here. If you like, walk up the grassy bank of the Belington reservoir dam for a better view. Welcome to Camp Laurel Hill.

Fun Fact: During the action here, George Rodgers of the Seventh Indiana Infantry taunted enemy skirmishers by reading aloud a fictitious newspaper account of the death of Confederate President Jefferson Davis.

Camp Laurel Hill—most of the earthworks here were later filled in by Federal troops.

Stop 4—Camp Laurel Hill: Gateway to the Northwest

General Robert Garnett's Confederate army occupied this mountain pass on June 16, 1861. The pass at Laurel Hill—along with one at Rich Mountain, twenty miles south—were the keys to Garnett's defense of Western Virginia. He called them the "gates to the northwestern country."

Approximately 4,000 Confederates camped here. General Garnett's headquarters tent was pitched beneath a maple tree nearby. Confederate soldiers dug rifle pits and artillery emplacements on the cleared heights to the west. Timber was felled to barricade both flanks. The goal: to block a Federal advance on the Beverly-Fairmont Turnpike.

"We are situated on hills [on] both sides of the turnpike as that is the only way the enemy can advance on us."—**Clayton Wilson, 1ˢᵗ Georgia Infantry**

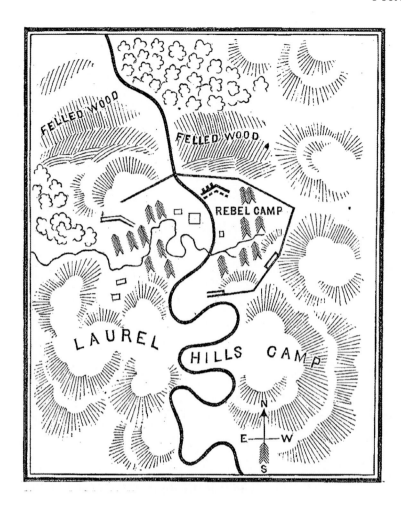

*Map of Camp
Laurel Hill by
Union spy William
B. Fletcher*

*"On a bench extending from the side of
Laurel Hill, they have constructed a bomb-
proof redoubt of logs, and on the left flank
they have three long rifle pits...Four old-
fashioned cannon, commanding the front
and side approaches, are mounted in the
fort."*—**New York Tribune, July 9, 1861**

At Camp Laurel Hill, the Confederate
army consisted of Virginians from
around the state. Among them was the
Thirty-first Virginia Infantry, made up of
volunteers from Western Virginia. Also in
this camp was the First Georgia Infantry.

The Georgians, more than one thousand
strong, boasted a fife-and-drum corps and
were brilliantly uniformed and equipped.
One Virginia colonel thought they came
"rather for a gay holiday than for real
war."

By July 10, the Federals planted
artillery within range of this camp and
opened fire. Isaac Hermann of the First
Georgia Infantry was on guard duty here
during the sporadic bombardment. By
watching for the smoke of the cannons,
Hermann discovered the sound took eight
seconds to reach him, followed about four

seconds later by arrival of the ball—time enough to duck into the trenches!

"They shot cannon balls, case shot and canister at us for near ten hours. We were sheltered from them, however, by the large trees in the woods."—James Hall, 31st Virginia Infantry

On the evening of July 11, General Garnett received chilling news: the Confederates had been defeated at Rich Mountain, twenty miles south. Garnett's position at Laurel Hill was now untenable. With the Federals under Morris testing his front, and those under McClellan threatening his rear, he had to evacuate.

Late that night, sullen soldiers and heavily laden Confederate wagons crossed over the mountain bearing south. Tents were left standing and campfires burning to deceive the enemy. The sun was high at Laurel Hill on July 12 before General Morris learned that Garnett was gone.

Note: *County Route 15 (the Beverly-Fairmont Turnpike) continues over the mountain crest, descends the east slope and eventually leads to Elkins. The road is rough and not recommended! Turn around, return to Camp Laurel Hill, and follow the directions below.*

Directions: Retrace your route to Belington and US 250. Turn left on US 250 and proceed south toward Elkins. Proceed to a stoplight at US 33/250 (50.8 miles), and turn left on the divided highway. Continue east, past the Elkins (Downtown) exit to the end of the freeway (62.4 miles). Follow the ramp to US 219 North (62.9 miles), turn right and continue north to Parsons.

☞ **Fun Fact:** The First Georgia Infantry dazzled all at Camp Laurel Hill with their finery. As the word spread, country folk came out on Sundays to gawk at their fancy uniforms, body servants, and silver dinner settings!

Confederate Retreat From Laurel Hill

On the night of July 11, 1861, the Confederate army under General Robert Garnett fled Camp Laurel Hill, tracing the Beverly-Fairmont Turnpike to Beverly, twelve miles south. Nearing Beverly at daybreak on July 12, Garnett's scouts reported that Federal troops held the town. This report was false, but Garnett did not know it. The troops seen in Beverly were actually Confederates fleeing from Rich Mountain!

To General Garnett, the news was grim. His Confederates were now cut off from retreat on the Staunton-Parkersburg

Whitelaw Reid—this Cincinnati reporter was one of the first to bring the grim reality of war to Northern doorsteps.

Turnpike—their lifeline to the Shenandoah Valley. Garnett's last avenue of escape was a rough wagon road leading northeast through the mountains. Backtracking a few miles, the general led his army along the Leading Creek road toward Red House, Maryland. From there, they would turn south to Monterey, Virginia, a rugged detour of nearly 150 miles.

The Chase

Meanwhile, a Federal strike force had been dispatched from Laurel Hill in pursuit of General Garnett. One of the most dramatic chases of the Civil War was underway. Leading the chase was Capt. Henry W. Benham, a veteran regular army officer, and the top-ranked graduate of his

1837 West Point class. Capt. Benham's force consisted of 1,840 Ohio and Indiana foot soldiers.

Benham's "bloodhounds" were eager for the chase. The Confederates had a twelve-hour lead, but were easily tracked. Deep mud marked their line of retreat, and the route was littered with camp equipment cast off to lighten the wagons.

Slowed by timber barricades and a pitiless rain, Benham's Federals reached the hamlet of New Interest (present-day Kerens) on the evening of July 12. Here the track of fleeing Rebels turned east on a mountain trace.

Directions:

Option 1. Continue north on US 219 to Parsons. Although this two-lane highway is curvy, it bypasses the narrow roads of the historic chase along Shavers Fork (Option 2).

Option 2. Follow the actual route of the chase by the Shavers Fork river route. Note: This route is a narrow, one-lane, paved road that winds over Pheasant Mountain to Shavers Fork of Cheat River, six miles east. Extreme caution should be used when traveling this road. Drive defensively!

If you choose Option 1: Continue north on US 219 to **Stop 6—A Wilderness Ordeal.** You will skip Stop 5, but read the narrative below.

Kalars Ford of Shavers Fork—this rocky crossing greatly slowed the Confederates fleeing from Laurel Hill.

Directions for Option 1: Continuing on US 219, turn right on Shavers Fork Road, (County Route North 39) (74.9 miles), bear left and follow the road for .2 miles to Riverview Chapel. Turn left into the chapel parking lot and park near the large brick sign. Note the fine view of Shavers Fork of Cheat River. **Go to Stop 6**.

Directions for Option 2: From US 219 at Kerens, turn right on Pheasant Mountain Road. (County Route 3) (63.9 miles). Proceed east along this one-lane paved route, climbing the wooded foothills of Pheasant Mountain. At the crest of the ridge (65.7 miles), the road enters Tucker County and becomes County Route 47 (no signs are present). Continue down the twisting mountain road past Irons Chapel on your right (69.1 miles). Bear left on Shavers Fork Road (County Route 39 North) and park on the right at a small pull-off in a turn (69.5 miles). Note the guardrails and large yellow turn arrows. Leave your vehicle, watch for traffic, and walk north beside the guardrails for about thirty feet. Look south to view Shavers Fork of Cheat River. **Do not step over the guardrails for any reason!**

*"That road defies description. Part of the time it ran through lanes so narrow that a horseman could not pass on either side of the wagon train; then it wound through mountain gullies where the wheels of the wagon would be on the sides of the opposite hills, while beneath rushed a stream of water."—***Whitelaw Reid**, **Cincinnati Gazette**

Stop 5—Kalars Ford: Raucous Retreat to Shavers Fork

The vanguard of General Garnett's retreating army reached Kalars Ford by the evening of July 12th. Approximately 3,500 Confederates and a large train of wagons were strung out along the rugged road for more than two miles.

Confederate axmen felled trees to block the road, but Capt. Benham's Federals were closing. At every step, the mud grew deeper. Soldiers slipped and staggered like drunken men in the mire. Discarded equipment cluttered the roadway. Wrecked wagons hung upside-down in the trees above dizzying precipices.

*"It was no longer the retreat of an army; it appeared the pell mell rout of a mob. The destruction of property was enormous. Fine, heavy duck tents, and elegant blankets, far better than the best of ours lay in the road and were trampled by the infantry and ground into the mud by the wagons."—****Whitelaw Reid**, **Cincinnati Gazette***

Here, on the morning of July 13, General Garnett's Confederate army crossed the rocky riverbed of Shavers Fork of Cheat River. The crossing, known as Kalars Ford, can still be seen: look 100 yards upstream (south); note the shallows and old roadbed entering the river below a residence. That roadbed, Garnett's escape route, crossed Shavers Fork a number of times as it wound downstream (north) toward St. George.

General Garnett's army and wagon train took most of the morning of July 13 to cross the rain-swollen river here at Kalars Ford. By noon, the pursuing Federal troops under Capt. Benham had reached the crossing. Garnett's wagon train was visible in a large field near the next ford, about one mile north. Plunging into the cold, deep water, Benham's bloodhounds broke into a chorus of the hymn "On Jordan's Stormy Banks I Stand."

Now the chase became a running fight.

*"The river was reached and forded again and again—water ice cold and in depth from thigh to waist; the advance guard overtook what they supposed to be the rear-guard of the enemy."—****Orville Thomson, 7th Indiana Infantry***

☞ **Fun Fact:** While crossing the river here, a Federal captain stood in midstream with sword in hand, ordering his troops to "dress to the right." As he slipped and toppled into the water, the men snickered: "dress up, Captain, dress up!"

35

Battle of Corricks Ford, July 13, 1861—Federals in the foreground strike the Confederate rearguard in a rainstorm.

Directions for Option 2: Return to your vehicle. County Route 39, a narrow, one-lane paved road, overlooks the route of the chase along the banks of Shavers Fork. Continue north on County Route 39 to Riverview Chapel on the right (71.8 miles). Turn right into the chapel parking lot, and park near the large brick sign. Note the fine view of the river.

Stop 6—A Wilderness Ordeal

Garnett's Confederates fled downstream, past the location of modern-day Riverview Chapel, with Benham's bloodhounds at their heels. To slow the pursuit, the First Georgia Infantry was posted above the second ford (out of view upstream). Small arms and Federal artillery

fire echoed across the valley. Many of the Georgians were soon cut off by the enemy, and scampered up the steep mountain on the eastern horizon into wild and uninhabited country.

Lost and famished, they blundered through immense laurel thickets for four terrifying days, peeling birch bark from the trees for sustenance. A trapper finally found the starving Georgians and led them to safety. The setting of their ordeal is today called the Otter Creek Wilderness!

The remainder of General Garnett's army fled down this scenic valley, greatly slowed by their wagons. At the next river crossing, Corricks Ford, the Confederates were forced to make a stand.

Map of Confederate Retreat from Laurel Hill and Battle of Corricks Ford

→ UNION
⇨ CONFEDERATE

(NOT TO SCALE)

Directions (Options 1 and 2): Leave the Riverview Chapel parking lot and continue north for .2 miles to the intersection of County Route 39 and US 219 at Porterwood (Option 1: 75.3 miles, Option 2: 72 miles). Turn right on US 219. Proceed north, past the large Kingsford Charcoal plant on your right (Option 1: 76.4 miles, Option 2: 73.1 miles). There is no stop here.

Battle of Corricks Ford: July 13, 1861

In 1861, two Shavers Fork river crossings were known as Corricks Ford. The first, or upstream crossing was here, near the present-day Kingsford Charcoal plant. The river ran deep at Corricks Ford, and General Garnett's wagons stalled in the rocky bed. In desperation, Garnett posted his rearguard on a bluff overlooking the ford: Colonel William B. Taliaferro's 23rd Virginia Infantry, and three guns of the Danville (VA) Artillery.

Members of the 14th Ohio Infantry, dressed in captured Confederate overcoats, soon charged the riverbank. A mighty

37

"Rebel yell" rang out, followed by a blaze of light as the defenders opened fire on the Federals below.

"Cheat River ran red with their blood." — **James Hall, 31ˢᵗ Virginia Infantry**

Federal troops crowded the riverbank, supported by a gun of the Cleveland Light Artillery. Cannonballs shrieked across the swollen ford. Bullets filled the air, "hissing like venomous serpents." The 7ᵗʰ Indiana Infantry attempted to scale the bluff, then waded downstream beneath the fire of both armies to cut off the Confederates. Colonel Taliaferro's Virginians were forced to retreat. The angry skirmish lasted about thirty minutes. Most of the Confederate wagons were abandoned at this crossing, still laden with fine officers' trunks and munitions of war.

"The first object that caught the eye was a large iron rifled cannon (a 6-pounder) which they had left in their precipitate flight. The star spangled banner of one of our regiments floated over. Around was a sickening sight. Along the brink of that bluff lay ten bodies, stiffening in their own gore, in every contortion which their death anguish had produced. Others were gasping in their last agonies...Never before had I so ghastly a realization of the horrid nature of this fraternal struggle."—**Whitelaw Reid, Cincinnati Gazette**

Captain Henry W. Benham—he led the chase that killed General Garnett and routed Confederates at Corricks Ford.

☞ **Fun Fact:** During the action at Corricks Ford, Sergeant William Copp, "fighting parson" of the Ninth Indiana Infantry, fired with perfect coolness and a steady aim. After every shot, he took down his gun and declared: 'And may the Lord have mercy on your soul!'"

BATTLE OF CARRICK'S FORD, WESTERN VIRGINIA—DISCOVERY OF THE BODY OF GENERAL GARNETT, BY MAJOR GORDON AND COLONEL DUMONT, AFTER THE BATTLE.

Union soldiers discover the body of Confederate General Garnett at Corricks Ford. Garnett was the first general killed in the Civil War.

Directions: Continue north on US 219. You may note a large stone monument and highway marker for Corricks Ford on the right side of the highway. **Do not stop here—it is unsafe.** Continue north to the Allegheny Highlands Trailhead (Kohler Street) (Option 1: 77.7 miles, Option 2: 74.4 miles). Turn right and park adjacent to the Allegheny Highlands Trail, a converted railroad grade. From here, you can walk the paved trail south for approximately .3 miles to the scene of General Garnett's death. You will pass interpretive signs along the way. Continue south to a point where you can view the river's edge. An interpretive sign marks the site of General Garnett's death.

Stop 7—Corricks Ford: Death of General Garnett

Here, on the banks of Shavers Fork, was the second Corricks Ford. At the sound of battle from the first crossing, one half mile south, General Robert Garnett left the vanguard of his retreating Confederate army and rode back to this ford. Mortified by the chaos he found at every turn, Garnett determined to make a last-ditch stand.

Selecting ten riflemen from the "Richmond Sharpshooters," 23rd Virginia Infantry, he posted them behind driftwood on the near bank of the stream. Federal skirmishers soon appeared. Bullets zipped across the ford.

General Garnett sat prominently on horseback at the river's edge. Aides pleaded for him to retreat. Turning in his saddle to give an order, Garnett was struck square

39

in the back by a ball and toppled to the riverbank. Federal skirmishers splashed across the ford and found him among the wildflowers, his muscles making "their last convulsive twitch."

"I have myself but little doubt that [General Garnett] returned in the expectation or hope of losing his life in mortification at this disastrous rout." —Capt. Henry W. Benham, U.S.A.

The fallen general wore a black overcoat and a uniform of blue broadcloth. Beside him lay a dead Confederate rifleman, with girlish locks of blonde. Union Major Jonathan Gordon placed guards over the bodies. Northern soldiers filed past in awed silence: Robert Garnett was the first general officer killed in the Civil War.

General Garnett's body was carried to the William Corrick house, overlooking this ford. Captain Benham's pursuit of the Confederates ended here. An attempt to cut them off via the B & O Railroad near Red House, Maryland, failed.

As word of General Garnett's death spread through the Southern ranks, his army fled in disarray. The rabble extended for miles. Citizens told of ravenous soldiers who caught poultry from barnyards as they passed, and, tearing off the feathers, devoured them raw. Days later, the remnant of Garnett's Confederate army reached safety in the Highland County village of Monterey, Virginia.

"We have suffered awfully. Not many men were killed by the enemy, but there are hundreds missing.... What is left of this army will not be fit for service in a month." —Colonel James N. Ramsey, 1st Georgia Infantry*

☞   **Fun Fact:** In one of the Civil War's great ironies, Union Major John Love identified the body of Confederate General Garnett at Corricks Ford. The two had been West Point roommates!

> **Directions:** Return to US 219. Check out the shops and restaurants of Parsons. If time allows, visit scenic Blackwater Falls and Canaan Valley State Parks. Return to Elkins.

Lodging, shops and restaurants await in Elkins to refresh you for the adventures on **DAY 2.**

END OF DAY 1

DAY 2

CAMP GARNETT, BATTLE OF RICH MOUNTAIN, BEVERLY AND HUTTONSVILLE, CAMP ELKWATER, MINGO FLATS, VALLEY MOUNTAIN AND BIG SPRING

George B. McClellan, a Union general full of promise.

Preview: Day 2 traces the Federal march to Camp Garnett and a climactic battle on the crest of Rich Mountain. The Federals emerge victorious. General George McClellan's performance is less than stellar, but dazzling telegrams to the War Department make him an instant sensation, the North's first battlefield hero! On July 22, 1861—after a stunning Federal defeat at Manassas—McClellan is summoned to Washington. President Lincoln has called him to save the Union.

The "Army of Occupation" in Western Virginia, now led by General William Rosecrans, goes on the defensive. Federal troops fortify key turnpike crossings atop Cheat Mountain, and in the upper Tygart Valley at Elkwater, to keep the Rebels out.

Meanwhile, the Confederate Army of the Northwest grows stronger. General W.W. Loring takes command as Rebel reinforcements pour across the Alleghenies. The Confederates have reason for optimism; General Robert E. Lee has arrived to "strike a decisive blow." Rugged mountain scenery, historic antebellum homes, shopping, and outdoor adventure await you on **DAY 2**.

Lieutenant Colonel John Beatty, 3rd Ohio Infantry

41

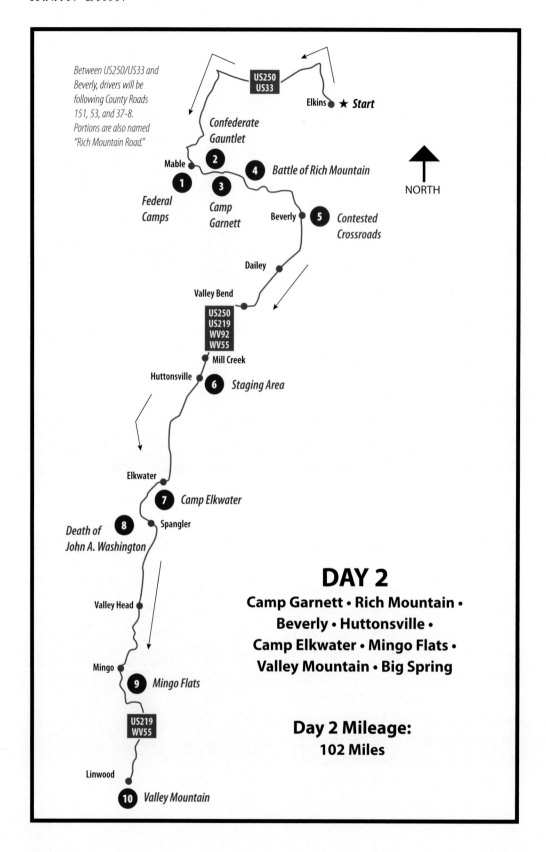

Between US250/US33 and Beverly, drivers will be following County Roads 151, 53, and 37-8. Portions are also named "Rich Mountain Road."

US250
US33

Elkins ★ *Start*

Confederate Gauntlet

NORTH

Mable
1
2
3
4 *Battle of Rich Mountain*

Federal Camps

Camp Garnett

Beverly
5 *Contested Crossroads*

Dailey

Valley Bend

US250
US219
WV92
WV55

Mill Creek
Huttonsville
6 *Staging Area*

Elkwater
7 *Camp Elkwater*

8 Spangler

Death of John A. Washington

DAY 2
Camp Garnett • Rich Mountain • Beverly • Huttonsville • Camp Elkwater • Mingo Flats • Valley Mountain • Big Spring

Valley Head

Mingo
9 *Mingo Flats*

US219
WV55

Day 2 Mileage:
102 Miles

Linwood
10 *Valley Mountain*

Directions: From the "Iron Horse" statue in Elkins, proceed north as on Day 1: go through one light, and get in the right lane. Turn right at the second light (0.1 mile) and continue on US 33 West (marked "To I-79") and turn left (2.0 miles) to enter the divided highway, US 33. Exit the highway left at County Route 151 (7.7 miles), following the signs for Rich Mountain Battlefield. Additional Battlefield signs are present at each turn along this route. Within one half mile, you will bear right and cross a large bridge over Tygart Valley River. Turn left on County Route 53 (8.5 miles) (Coalton-Pumpkintown Road), driving through Norton and past the village of Coalton. Turn left on the Mabie-Cassity Road (County Route 37/8) (12.9 miles) and proceed until you see the Mabie Post Office, on the right (14.0 miles). Turn right into the post office parking lot, and walk to the flagpole.

Stop 1—Federal Camps on Roaring Creek Flats

On the afternoon of July 9, General George McClellan led three Federal brigades (approx. 6,000 men) from Buckhannon along the Staunton-Parkersburg Turnpike to Roaring Creek Flats. "We came over the hills with all the pomp and circumstance of glorious war," wrote Lt. Colonel John Beatty of Ohio, "infantry, cavalry, artillery, and hundreds of army wagons; the whole stretching along the mountain road for miles."

Signs of impending conflict lay ahead. Confederate bayonets glistened from the heights of Rich Mountain, visible to the east. The bridge over Roaring Creek, less than one half mile ahead, was smoldering in ruins. Anxious Federal troops pitched their tents in the meadows nearby.

"These mountain passes must be ugly things to go through when in possession of an enemy; our boys look forward, however, to a day of battle as one of rare sport...I endeavor to picture to myself all its terrors, so that I may not be surprised and dumbfounded when the shock comes."—**Lieutenant Colonel John Beatty, 3rd Ohio Infantry**

General McClellan knew almost nothing of the Confederate defenses on Rich Mountain, less than two miles east. "The enemy are in sight," he wrote to wife Nelly on July 10, "& I am sending out a strong armed reconnaissance to feel him & see what he is. I have been looking at the camps with my glass—they are strongly entrenched..."

The Ninth Ohio Infantry, a crack regiment of German immigrants, led that reconnaissance. The "Bully Dutch" drove Confederate skirmishers to their trenches, but were stopped by a withering fire from the heights.

McClellan's tone became less confident. "I realize now the dreadful responsibility on me—the lives of my men—the reputation of the country & the success of our cause," he confided in a note to Nelly. "I shall feel my way & be very cautious...." Unwilling to send his green troops against those defenses, McClellan sought a way around them.

Union Major General George McClellan and his charming wife 'Nelly.'

> **Directions:** From **Stop 1**, continue south on County Route 37/8. In a sharp curve, turn left on the Rich Mountain Road (15.0 miles) (CR 37/8), a one-lane paved roadway that traces the historic Staunton-Parkersburg Turnpike. You are now in the area through which the armies skirmished during McClellan's July 10 reconnaissance. From this intersection, continue ahead. You will bear right at "Field of Fire" Park (15.6 miles) and continue to a parking area in a field on the left (15.9 miles). Turn left and park here, facing east. In the woods ahead are the remains of Confederate Camp Garnett.

Stop 2—A Confederate gauntlet

During June-July 1861, Confederate troops fortified the wooded ridges in your front. These defenses were named "Camp Garnett" in honor of the commanding general. Here foothills converge at the western base of Rich Mountain to form a narrow passage for the Staunton-Parkersburg Turnpike.

General Garnett believed this bottleneck would allow a fortified detachment to hold back "five times their number," and left 1,300 Confederate soldiers here for its defense. Confederates dug fortifications on the mountain spurs above you and felled timber along the turnpike for several hundred yards to the west, providing a field of fire for artillery—and a barrier to advancing troops. These bristling defenses created a gauntlet that Federal soldiers called the "Valley of Death."

Directions: From the parking area, turn left and continue a short distance along the Staunton-Parkersburg Turnpike (County Route 37/8) to another pull-off on the left side of the road beside a large oak tree (16.0 miles). Watch for traffic! Park here and walk down the short interpretive trail leading north to a wooden observation deck. In the woods to the north, directly across a small creek, can be seen remnants of the defenses of Confederate Camp Garnett.

Stop 3—Camp Garnett

The earthworks in the woods directly in your front are the remains of Camp Garnett. Approximately 1,300 Confederate troops were stationed here during June-July, 1861. All were Virginians, mostly of the 20th and 25th Virginia Infantry regiments. Lt. Colonel John Pegram of the 20th Virginia, a dashing West Point graduate, was the field commander.

To build their defenses, Confederate soldiers cut down trees on the hillsides and rolled them into place, shoveling dirt and rocks from a large trench (visible in front of you) to form an embankment. This earthwork protected them from enemy fire.

Additional works were constructed on the long ridge south of the vital turn-

The Staunton-Parkersburg Turnpike:

This Virginia toll road, completed by 1847, linked the Shenandoah Valley town of Staunton with Parkersburg on the Ohio River. Claudius Crozet, a distinguished French engineer, surveyed portions of the rugged route with chain and compass. This stone-based turnpike was a major east-west corridor across Western Virginia, and the focus of armies North and South. You will follow its route, more or less, for much of **Days 2** and **3.**

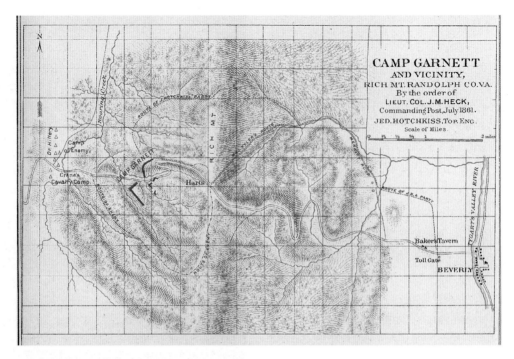

Map of Camp Garnett drawn by Jed Hotchkiss

Trenches of Camp Garnett

David Hart, the Rich Mountain guide.

pike. Four cannons anchored the defense of Camp Garnett—small 6-pounder smoothbores that fired shot and shell.

The Confederates pitched tents on the slopes behind their works. Among the notables here was Jed Hotchkiss, a civilian surveyor who mapped the ground around Camp Garnett, launching a career that would make him the most famous map-maker of the Civil War. Also camped here were students from Hampden-Sydney College, the "Hampden-Sydney Boys" (Co. G, 20th Virginia Infantry), led by Rev. Dr. John Atkinson, their president!

A Flank March

On the evening of July 10, one of McClellan's brigadiers—General William S. Rosecrans—interviewed a twenty-two year old civilian named David Hart. Young Hart offered to guide the Federals up a rugged path that led south of the Confederate works to his father's home on the turnpike at the crest of Rich Mountain.

General Rosecrans ushered Hart to McClellan's headquarters. There he proposed an assault. With this young guide, Rosecrans would lead a brigade of Federal troops on a flank march to the Hart house the next morning, gaining the turnpike nearly two miles in rear of the Confederate works. The route was barely passable for horsemen, and too rugged for artillery.

After gaining the turnpike on the mountaintop, Rosecrans was to fall upon

Jed Hotchkiss, mapmaker for Stonewall Jackson and others, began his career at Rich Mountain.

the rear of Camp Garnett. The sound of his musketry would signal General McClellan to launch a frontal attack—trapping the Confederates. McClellan approved the plan. By 5:00 a.m. on July 11 in a light rain, Rosecrans' brigade was silently underway.

Directions: Return to your vehicle and continue east on the Rich Mountain Road. (County Route 37/8), following the gravel roadway. This is the old grade of the Staunton-Parkersburg Turnpike, a

National Scenic Byway. As the turnpike climbs, note the switchback turns! The road has been widened for the passing of modern traffic. Continue upward until you reach a parking area on the right at the top of Rich Mountain (17.8 miles). Note the Rich Mountain Battlefield signs. Turn right and park here. Welcome to Rich Mountain Battlefield—one of the most important Civil War sites in America.

Lieutenant Colonel John Pegram, dashing commander of Confederate forces at Rich Mountain.

Stop 4a—Battle of Rich Mountain: July 11, 1861

(4a) From the parking area, take a short walk to the first interpretive sign. In 1861, this was the homestead of Joseph Hart and family. The Staunton-Parkersburg Turnpike crossed this gap on Rich Mountain at an elevation of more than 3,000 feet. The side roads did not exist at that time.

Union General Rosecrans' five-mile flank march to the summit on July 11 was more difficult than imagined. His 1,900-man brigade, led by David Hart and the intrepid Frederick Lander, fell hours behind schedule. They finally reached the mountaintop about one mile south of this point at 2:30 P.M., when enemy pickets opened fire.

The shooting alerted Confederates here at the gap. Earlier that morning, a captured Union courier had revealed that

Union General William S. Rosecrans. He planned and won the battle of Rich Mountain.

a flank march was underway. Colonel Pegram had placed 310 Confederates at the Hart farm (members of the 20th and 25th Virginia Infantry regiments, a detachment of the Churchville Cavalry, and a single cannon of the Lynchburg Lee Battery), under the command of Captain Julius DeLagnel. His orders were to defend this post "to the last extremity."

At the sound of gunfire, the Southerners leaped behind large boulders and log works on the opposite side of the turnpike. Rain hammered down as Federal troops appeared, descending the wooded slope from the south. General Rosecrans' force consisted of the 8th, 10th, and 13th Indiana, and 19th Ohio Infantry regiments. The lone Confederate cannon, a 6-pounder located beside a log stable just across the turnpike, opened in a fury.

Note on the Rich Mountain Battlefield Foundation

Since its inception in 1991, the Rich Mountain Battlefield Foundation has acquired and protected more than 400 acres of this important site. For information on programs and membership, contact Rich Mountain Battlefield Foundation at 304-637-RICH or www.richmountain.org.

"The whole earth seemed to shake." —David Hart, Rich Mountain guide

Confederate gunners pounded the Union line with shot and shell. Thrown into confusion, Federal soldiers huddled behind rocks and trees. David Hart, terrified by that rapid fire, thought the

49

The Hart house, scene of battle on the crest of Rich Mountain.

Confederates had twenty-five or thirty cannons!

After much delay, General Rosecrans reformed his green troops and brought them into line. "Charge bayonets," he roared, and the long line of Federals swept downhill and across the turnpike with a yell.

Stop 4b: The Hart House

(4b) From this point, walk east across the dirt lane through the grass to an interpretive sign at the brow of Rich Mountain. Here stood the Joseph Hart house, built in 1855. This two-story home was caught in the crossfire of battle and sustained much damage. After the battle,

the house was used as a hospital. It burned in 1940, with bloodstains still visible on the floors.

Joseph Hart, a loyal Unionist, was descended from a signer of the Declaration of Independence. His son David, the Rich Mountain guide, later joined the 10th Indiana Infantry as commissary sergeant, and died of illness near Nashville, Tennessee in 1862.

"The bloody-handed surgeons, with... bandages, saws, scalpels, probes, and bullet forceps, were busy bandaging and dressing what could be saved, and amputating hopelessly shattered and lacerated limbs." — **R. A. Riley, 8th Indiana Infantry**

Fun Fact: The headless ghost of a Confederate soldier reportedly haunted the Hart House until it burned. This apparition disturbed persons sleeping in an upstairs bedroom, often at the stroke of midnight!

Stop 4c: Struggle in the Stable Yard

(4c) Leaving the Hart house, cross the Staunton-Parkersburg Turnpike (watch for traffic!) and walk to the interpretive trail at the "stable yard" sign. You are now standing at the heart of the Confederate defenses during the Battle of Rich Mountain.

Immediately to your right (east) stood a log stable, and the lone Confederate cannon. Log fieldworks protected Virginians on the mountainside just above, which was then partly cleared. The large sandstone boulders nearby provided cover for many defenders. Veterans of the battle carved inscriptions on some of these boulders to mark the spots where Confederate soldiers fought and died. These inscriptions are now worn and difficult to read. **Please don't try to highlight them with pigment for photographs, or take rubbings—it will only speed their decay!**

"At one large rock about 30 feet long, behind which the enemy had been concealed, there laid, piled upon and across one another, sixteen men, every one of which was shot through the brain." —**R. A. Riley, 8ᵗʰ Indiana Infantry**

Inscribed Boulder: "Clay Jackson, shot and killed here in 1861."

Henry Clay Jackson was a member of the "Upshur Grays," 25th Virginia Infantry. Before the battle, young Jackson boasted he would "kill a damn Yankee and cut out his heart and roast it." He was shot in the neck and died without having fired a shot!

The outnumbered Confederates at the stable yard held the Federals back for more than two hours before they were overrun. Near the end of the fight, Lieutenant Colonel Pegram arrived with reinforcements and a second cannon, but it was too little, too late. Federal troops swarmed into the stable yard, bayoneted the defenders, and captured both guns. Lieutenant Colonel Pegram fled back down the mountain to Camp Garnett.

The Battle of Rich Mountain was over. One of every four Confederates at the

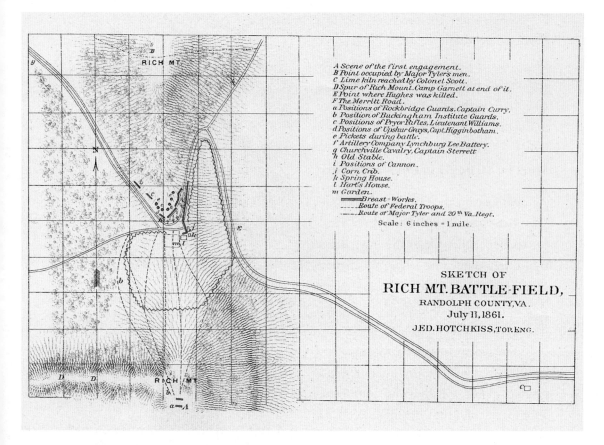

Map of Rich Mountain Battlefield by Jed
Hotchkiss

Hart farm had been killed or wounded. Most were buried in a large trench near the road. After the war, these Confederates were reburied in a mass grave on Mt. Iser in Beverly. The Federal losses were officially 12 killed and 62 wounded. Most were buried on the battlefield, and later moved to National Cemeteries.

*"It was a bloody affair when we consider the number engaged."—**Lt. Orlando Poe, U.S.A.***

Aftermath

On Rich Mountain, the victorious General Rosecrans found himself in a quandary. It was 6:00 p.m.—too late to fall upon the Confederates at Camp Garnett, nearly two miles below. General McClellan had not launched a frontal attack. McClellan later explained his inactivity by describing the firing on the mountaintop as "distant and stationary." He feared that Rosecrans had been defeated!

Battle of Rich Mountain, July 11, 1861—a small Union victory with huge repercussions

At daylight on July 12, after a wet, sleepless night, Rosecrans' men moved cautiously down the turnpike and found Camp Garnett abandoned. The Confederates had fled through the woods under cover of darkness, leaving all of their ordnance and equipment, plus 70 sick and wounded soldiers. A messenger rode through the deserted camp to inform General McClellan of the victory.

McClellan promptly ordered a march on Beverly. His raw recruits were stunned by the carnage they witnessed on passing the Rich Mountain battlefield.

"They had about twenty five [dead Confederates] thrown into a ditch, and of all the horrid sights I ever looked upon this was the most horrid. They were thrown in without any regard to order or the usual rites of scripture, some with shattered skulls, mangled limbs or ghastly bayonet wounds."— **C. R. Boyce, 3rd Ohio Infantry**

Fellow Americans had become deadly foes.

West Virginia and Regional History Collection, WVU Libraries

Stonewall's Sister

Laura Jackson Arnold of Beverly was a staunch Unionist, and the sister of Confederate General Thomas "Stonewall" Jackson. She parted with her brother after the outbreak of war and became a celebrated nurse. It was said that Laura could "nurse Federal soldiers as fast as her brother could wound them."

Directions: Return to your car. Descend Rich Mountain, continuing east on the Staunton-Parkersburg Turnpike (County Route 37/8), through a three-way stop (21.3 miles) and toward the town of Beverly. Upon crossing the bridge over Tygart Valley River, note the large home immediately to your right—the Lemuel Chenoweth House and Museum. Here lived Lemuel Chenoweth, a gifted craftsman who built the famous Philippi covered bridge.

Continue to the intersection of US 219/250 (22.7 miles). The two-story red brick house on the corner on your left was the home of Laura Jackson Arnold, sister of Confederate General Thomas "Stonewall" Jackson.

Turn right on US 219/250. Continue less than .1 mile and turn left on the Files Creek Road. Take the first right, and park at the Beverly Heritage Center.

Stop 5—Beverly: Contested Crossroads

Established in 1790, Beverly was the seat of Randolph County until 1899, and an important center of commerce. The town is situated at a key crossroads. Here the Staunton-Parkersburg Turnpike intersects the Beverly-Fairmont Turnpike, a route that led north to the Baltimore and Ohio Railroad.

Beverly's setting made it a focal point of conflict throughout the Civil War. On July 12, 1861, one day after the Battle of Rich Mountain, Union General McClellan's army seized the town. Young

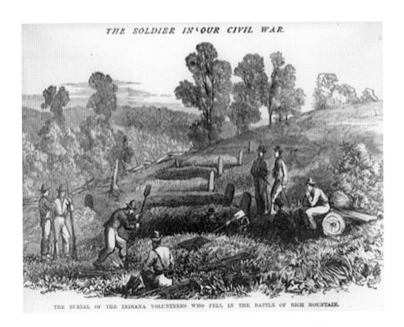

Burial of the Indiana Volunteers on Rich Mountain

Thomas Arnold, nephew of Stonewall Jackson, recalled their arrival:

"[N]*umerous regiments in handsome uniform, with banners flying keeping exact step to the music of the fine Regimental Bands, playing the beautiful March from the Opera, 'Norma.*'"

By occupying Beverly, McClellan cut Confederate General Garnett's supply line and blocked his retreat on the Staunton-Parkersburg pike.

The Bushrod Crawford House: McClellan's Headquarters

The Bushrod Crawford house (ca. 1850), now part of the Beverly Heritage Center, became General McClellan's headquarters. The act was symbolic; Crawford, an outspoken secessionist, had already fled south. In Beverly, on the morning of July 13, McClellan received

George B. McClellan, the "Young Napoleon." His victories and dramatic telegrams from Western Virginia made him the North's first battlefield hero.

*The Beverly
Heritage Center*

the surrender of Lieutenant Colonel Pegram and nearly 600 Confederates who had fled Camp Garnett. One day later, he learned of the rout of Confederate forces retreating from Laurel Hill, and of General Garnett's death at Corricks Ford. It was electrifying news!

Now McClellan unveiled his secret weapon—the telegraph. Coils of telegraph wire had trailed his advance, technology never before used by an American army in the field. A flurry of dramatic messages danced over the wires from McClellan's headquarters to the War Department in Washington:

"Garnett's forces routed...his army demoralized...have taken all his guns, a very large amount of wagons, tents &c.—everything he had....Garnett killed....We have annihilated the enemy in Western Virginia....Our success is complete & secession is killed in this country."

These were the first Union victories of the war, and the news sparked a sensation. The young general's less than stellar behavior on the battlefield was forgotten.

Thanks to his dazzling telegrams, George McClellan became the North's first battlefield hero!

On July 21, Federal forces suffered a stunning defeat at Manassas, just outside of Washington D.C. A day later, the telegraph at Beverly tapped out more heady news:

PRESIDENT LINCOLN SUMMONED GENERAL MCCLELLAN TO WASHINGTON WITHOUT DELAY.

America's newest hero, the "Young Napoleon," was called to save the Union. By November 1861, without taking the field, McClellan became General-in-Chief—commander of all Union armies. *"I have made a very clean sweep of it—never was more complete success gained with smaller sacrifice of life."*—**Major General George McClellan, U.S.A**

☞ **Fun Fact:** Laura Jackson Arnold, sister of General "Stonewall" Jackson, nursed Confederate Lieutenant Colonel John Pegram in her Beverly home after his surrender. General McClellan, a classmate of Laura's brother at West Point, visited during Pegram's convalescence.

While in Beverly, be sure to visit the Beverly Heritage Center, the Randolph County Historical Society Museum, and other craft and tourist destinations in this charming historic village.

Directions: After your tour of historic Beverly, continue south on US 219/250 to the town of Huttonsville (34.2 miles). Here, US 219/250 splits. Turn right on US 219 and immediately turn left into an Exxon station. To the right is an interpretive sign for Huttonsville. Park here.

Stop 6—Huttonsville: Staging Area for the Blue and Gray

In June 1861, Huttonsville was a point of rendezvous for Confederates after the "Philippi Races." Confederate General Garnett made temporary headquarters in the town, reorganized the Army of the Northwest, and from this point advanced to fortify the mountain passes at Rich Mountain and Laurel Hill. Following Union General McClellan's success at Rich Mountain, Federal forces occupied Huttonsville.

This tiny hamlet was situated at the junction of two mountain roads. The Staunton-Parkersburg Turnpike (now US 250) led across Cheat Mountain to the Virginia Central Railroad at Staunton, almost one hundred rugged miles southeast. A secondary route, the Huttonsville-Huntersville Turnpike (now US 219) led due south from Huttonsville to Huntersville (former seat of Pocahontas County), then across the Alleghenies to the Virginia Central Railroad at Millboro.

On July 13, General McClellan advanced to Huttonsville, following the Staunton-Parkersburg Turnpike to the crest of Cheat Mountain the next day. Before he departed for Washington, McClellan ordered the erection of forts on both roads. His goal was to keep the Confederates out of Western Virginia.

By July 16, Federal troops dug in along the Staunton-Parkersburg pike on Cheat Mountain. Soon after, Federals began to fortify the narrow Tygart Valley seven miles south on the Huttonsville-Huntersville road at Elkwater.

Union General William Rosecrans now took command of the "Army of Occupation—Western Virginia." Anxious to support operations in the Kanawha Valley, 125 miles south, Rosecrans moved headquarters to the railroad at Clarksburg, and left General Joseph J. Reynolds in charge of the First Brigade at Huttonsville. General Reynolds (West Point class of 1843) could marshal some 9,000 men, less than half of McClellan's original force. With pressing needs in other theaters of war, Federal troops here now went on the defensive.

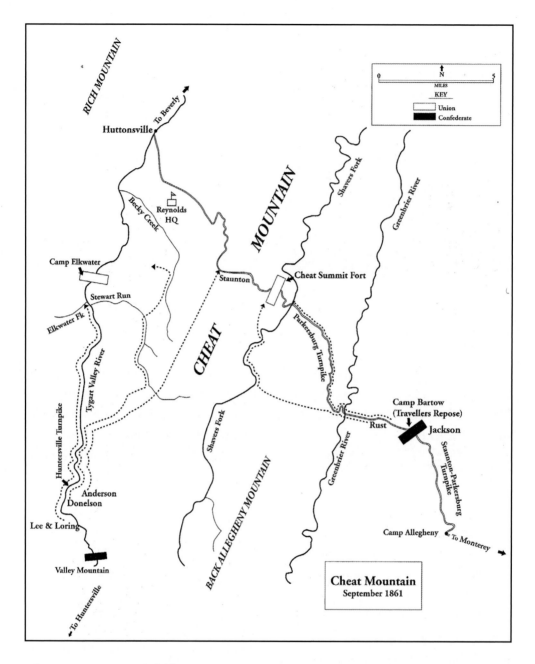

Cheat Mountain Battlefield

"Next we came to Huttonsville, a place
of not much fame—
    Two houses and a blacksmith shop is all
the place contains."
—**"Colonel" Coe, 32nd Ohio Infantry**

Meanwhile, the shattered Confederate Army of the Northwest reorganized in the village of Monterey, Virginia, fifty miles southeast on the Staunton-Parkersburg Turnpike. Fresh recruits marched across the Alleghenies to join them. By July 24, Confederate General William W. Loring, a battle-hardened veteran who had lost an arm in the Mexican War, took command.

General Loring made headquarters at Huntersville, fifty miles south. He soon marshaled nearly 11,000 troops. Hailing from various Southern states, these Confederates were drilled and ready to fight. There was renewed hope—General Robert E. Lee was riding west into the mountains. The First Campaign now entered a second phase.

General Joseph J. Reynolds, U.S.A.

Directions: Return to US 219, and continue south. On the right (41.0 miles), you will see interpretive signs for Camp Elkwater. Carefully pull off on the right at the driveway junction here. Please respect private property! Read the signs, and look back down the valley in the direction from which you came (north).

Stop 7—Camp Elkwater: Gateway to the Tygart Valley

"This is by far the pleasantest camp we have ever had…. My tent is on the bank of the

General William W. Loring, C.S.A.

*Valley river. The water, clear as crystal as it hurries on over the rocks, keeps up a continuous murmur."—**Lieutenant Colonel John Beatty, 3rd Ohio Infantry***

By late July 1861, Federal troops under General Joseph Reynolds began fortifying the narrow valley here at Elkwater. Camp Elkwater was erected to block Confederate incursions north along the Huttonsville-Huntersville Turnpike.

Camp Elkwater was built in tandem with a Federal fortress astride the Staunton-Parkersburg Turnpike on Cheat Mountain, seven miles east. The two camps were linked by the turnpikes at Huttonsville, and by bridle paths across the mountains.

The defensive works at Camp Elkwater are still preserved, marked by a line of trees crossing the Tygart Valley, visible by looking south from this point. Cannons were entrenched on spur ridges overlooking the valley. Colonel George D. Wagner of the Fifteenth Indiana Infantry led more than a thousand Federals in plying axe and spade on these fortifications.

*"Acres of forests, covering the hillside on either hand, were felled in the construction of breastworks and in clearing away what might prove a cover to the enemy."—**E. Hannaford, 6th Ohio Infantry***

On August 3, Confederate General Robert E. Lee reached General Loring's headquarters at Huntersville. Loring, now outranked, did not welcome Lee's effort to "coordinate" an assault. But when Lee

General Robert E. Lee, as seen in 1861.

advanced to Valley Mountain, only twelve miles south of Camp Elkwater, Loring was obliged to follow. By early September, despite prolonged rain and intense cold, the Confederates prepared to "strike a decisive blow."

Nearly 3,000 Federal troops were stationed here on September 12, 1861 when General Lee attacked. But Lee's assault on Cheat Mountain failed to come off. Hoping to fall upon the rear of Camp Elkwater, he rode to the crest of the ridge visible on the horizon northeast. On those heights, Lee joined General Daniel S. Donelson's Tennessee brigade, 1,600 strong. Donelson's Confederates had marched through the mountains for three days to reach this ridge. Exhausted and without food, their muskets and spirits drenched by a terrible storm on the night of September 11, they were utterly unable to strike. Later that day, at great risk, Lee guided them safely to the rear.

"I could see the enemy's tents on Valley River, at the point of the Huttonsville road just below me. It was a tempting sight. We waited for the attack on Cheat Mountain, which was to be the signal, till 10 A.M.; the men were cleaning their unserviceable arms. But the signal did not come."—**General Robert E. Lee**

For two days, Confederates skirmished with General Reynolds' army in front of Camp Elkwater. By September 14, failing to draw the Federals out of their defenses, Lee ordered his troops to withdraw.

Federal troops garrisoned Camp Elkwater until the spring of 1862. Many notables tented here, including future members of Congress, a Supreme Court justice, and future Presidents Rutherford B. Hayes and William McKinley.

☞ **Fun Fact:** On September 12, 1861, General Robert E. Lee was nearly captured by a squad of Federal cavalry near Camp Elkwater. Lee's aide Walter Taylor described the incident as "a very close call." How might Lee's capture have changed American history?

Rutherford B. Hayes, 23rd Ohio Infantry, a future president

Directions: Return to US 219 and continue south. Just before reaching the concrete bridge over Elkwater Fork, pull off in a small parking area on the left (42.5 miles). Look across US 219 to the west, up the little valley of Elkwater Fork.

Mount Vernon: Washington's Home and the Nation's Shrine

Lt. Col. John A. Washington, aide to General Lee.

Stop 8—Death of John Augustine Washington, C.S.A.

On September 13, 1861, General Lee ordered a reconnaissance of the Federal right flank at Elkwater. Leading this effort was the general's son, Major W.H.F. "Rooney" Lee. Lieutenant Colonel John Augustine Washington—a great-grandnephew of the first president and General Lee's aide-de-camp—joined young Lee's cavalry. Riding to a hilltop on the west, the group spied a mounted Federal soldier in the valley not far from where you stand.

Putting spurs to their horses, Washington and Lee galloped down this valley in a daring attempt to capture him. They did not know that scouts of the 17th Indiana Infantry were prowling the wooded slope to the northwest. As the two Confederates rode down this valley, Federal scouts fired at them from the hillside. Major Lee managed to escape, but Lieutenant Colonel Washington toppled from his bay charger.

Federal soldiers rushed to the fallen Washington. Pierced through the breast by three balls, he gasped for water, but died before it reached his lips. The finely dressed officer was carried to a nearby Federal outpost, where his identity was learned.

John Augustine Washington had sold the family estate, Mt. Vernon, prior to the war—a controversial move. Adding insult to the North, he then joined the Confederate army. Upon a smooth-barked beech tree (now gone) Federal soldiers carved his memorial:

"Under this tree, on the 13th of Sept., 1861, fell Col. John A. Washington, the degenerate descendant of the Father of his Country."

Washington's body was returned to the Confederates the next day under flag of truce. General Lee was deeply saddened. He had lost a valued aide, a tent mate, *and* a relative (Lee's wife was a Washington descendent). Lee wrote a painful letter of condolence to the eldest of Washington's seven young children, now orphans!

The death of John A. Washington ended Lee's offensive. With the loss of Lieutenant Colonel Washington and

'61 '65

LT. COL. JOHN AUGUSTINE WASHINGTON, C.S.A.
OF MOUNT VERNON,
AIDE-DE-CAMP TO GENERAL ROBERT E. LEE.
KILLED, ELKWATER, SEPTEMBER 13, 1861
BURIED IN ZION CHURCHYARD
CHARLES TOWN, WEST VIRGINIA.

General Garnett, the First Campaign had claimed half of Lee's original staff.

"My dear Miss Louisa, with a heart filled with grief, I have to communicate the saddest tidings which you have ever heard." — **General Robert E. Lee to the daughter of John Washington**

Monument to the death of John A. Washington at Elkwater.

☞ **Fun Fact:** John A. Washington's death and pedigree sparked much curiosity. The Federals doled out his belongings as war trophies. Union General Reynolds received Washington's field glass. A pair of gauntlets, a revolver, a large knife, spurs, and a powder flask went to members of the 17th Indiana Infantry. A second revolver was presented to the U.S. Secretary of War. There seemed to be general regret that Washington's horse had galloped away with his sword!

Directions: Return to your vehicle and continue south on US 219. Within 1/2 mile, you will cross a bridge over the headwaters of Tygart Valley River. The large fields near this crossing mark the position held by Confederate forces under Generals Lee and Loring during the offensive against Camp Elkwater. The Confederates abandoned Elkwater by September 14, 1861 and retreated south on the Huttonsville-Huntersville Turnpike. You are following their march along the modern road.

Continue south on US 219 until you reach the Mingo Flats Road (County Route 51) (52.3 miles). Turn right, and cautiously follow this one-lane paved route as it climbs a hill. Along the way, note an impressive bronze statue commemorating the "last Indian settlement in West Virginia" (52.5 miles). Mingo Indians reputedly lived nearby.

Proceed less than one mile to a statue of an armed Confederate soldier, on the left side of the road (53.3 miles). Find a safe pull-off and enter the fenced area enclosing the statue.

Stop 9—Mingo Flats: Confederate Staging Area

Members of Camp Pegram, United Confederate Veterans, erected this statue on July 23, 1913. It is dedicated "to the memory of the Confederate soldiers of Randolph County and vicinity. This includes all soldiers who died on Valley Mountain in 1861, while General Lee was encamped there."

The large fields nearby are known as "Mingo Flats." This locale was a staging area for Confederate troops under Generals Lee and Loring during their offensive in September 1861. From Mingo Flats, four Confederate brigades marched north along the turnpike and by rugged mountain paths to Camp Elkwater and Cheat Mountain.

A number of skirmishes took place between Mingo Flats and Camp Elkwater during the Confederate advance. John Worsham of the Twenty-First Virginia Infantry recalled seeing his first dead Yankee: "He made a lasting impression," Worsham wrote, "for he lay on the side of the road, his face upturned and a fresh pool of blood at his side, showing that his life had just passed away."

After the aborted assaults at Cheat Mountain and Elkwater, Confederate troops abandoned Mingo Flats and returned to Valley Mountain, three miles south.

☞ **Fun Fact:** During the advance near Mingo Flats, a brigade of Tennessee Confederates spied a "comely Virginia lass" at the window of a cabin. Each regiment in passing gave her a mighty "Rebel yell." The soldiers thought she was "highly gratified" by their demonstration!

Directions: Return to your vehicle and continue south on the Mingo Flats Road (County Route 51). The road soon rejoins US 219 (54.9 miles). Turn right at this intersection, and continue south on US 219. The winding road climbs until you reach the Randolph-Pocahontas County line at Mace. Immediately to your right (west) lies Valley Mountain.

Confederate monument at Mingo Flats. Erected by veterans in 1913 to commemorate those who served and died here in 1861.

General Robert E. Lee reached Valley Mountain on August 6, 1861, joining the 21st Virginia and 6th North Carolina Infantry regiments. His headquarters tent was reportedly pitched on the mountain crest at the county line. Lee shared a tent with aides John A. Washington and Walter Taylor. From these heights, Lee wrote to his wife Mary on August 9:

"We are on the dividing ridge.... In the valley north of us lies Huttonsville and Beverly, occupied by our invaders, & the

Rich Mountains west, the scene of our former disaster, & the Cheat Mountains east, their present stronghold, are in full view."

Directions: Continue south to the entrance of Snowshoe Mountain resort (58.9 miles). Turn left on WV 66, then immediately turn right into the parking lot adjacent to the Snowshoe Mountain entrance sign.

Stop 10—Valley Mountain and Big Spring: Confederate Camps of Measles and Mud

The area from this point to the Randolph-Pocahontas County line three miles north was often known as "Valley Mountain" in military correspondence. This place was named "Big Spring" for a large limestone spring at the base of the hill across US 219 to your west. Tradition holds that, for a time, General Lee's tent was pitched on the knoll above.

By August 1861, Confederate camps dotted the fields in this area. Regiments from Virginia, Tennessee, Georgia, and North Carolina marched here from the old county seat of Huntersville, twenty-five miles south. General Loring joined Lee at Valley Mountain by August 12, swelling forces on the "Huntersville line" to about 6,000 men. Additionally, 5,000 Southern troops held the Staunton-Parkersburg Turnpike on the "Monterey line" southeast of Cheat Mountain—giving the Confederates a combined strength of almost 11,000. They now outnumbered the Federals under General Joseph Reynolds, less than 9,000 strong.

The Confederates prepared to give battle. As mapmaker Jed Hotchkiss sketched the contested ground, General Lee scouted the Federal defenses. Lee spent many days in the saddle, often riding between the lines. He sought a route to flank the enemy, hoping to strike the fortress on Cheat Mountain by surprise.

But cruel weather intervened. The very heavens opened upon Lee's arrival. "We were camped on Valley Mountain 43 days," wrote a Virginian, "and it rained 37 days out of the number." One officer compared his camp to a "Tennessee hog pen." The muddy roads became nearly impassable. Unable to bring up supplies, the Confederates were placed on short rations.

"In all my experience of the war I never saw so much mud. It seemed to rain every day. It got to be a saying in our company that you must not hallo loud; for if you should, we would immediately have a hard shower. When some of the men on their return from picket had to shoot off their guns to get the load out, it brought on a regular flood."— **John Worsham, 21st Virginia Infantry**

Next came the cold. "The wind blows like winter," groused a soldier at Valley Mountain on August 16. "Ice was abundant yesterday morning, a large frost covering the ground." Even General Lee was stunned by this latest trick of nature: "The cold," he wrote, "has been greater than I could have conceived. In my winter clothing and buttoned up in my overcoat, I have still been cold."

Adding injury, an epidemic of measles, dysentery, and typhoid fever swept the Confederate camps. Sickness cut the army's strength in half. Rows of crude headstones soon covered these hills. Today, many unmarked graves remain beneath the sod.

By early September 1861, the sun reappeared, drying out the roads and

allowing the Confederates to advance. "Special Order No. 28" outlined a plan of attack. On September 9, the Confederates marched off to battle.

General R. E. Lee's headquarters at Valley Mountain

"*Let each man resolve to be victorious, and that the right of self-government, liberty and peace, shall in him find a defender.*"— **General Robert E. Lee**

Despite Lee's optimism, the assaults on Cheat Mountain and at Camp Elkwater failed miserably. Inexperienced troops and punishing weather doomed the ambitious plan. The Confederates suffered awfully. Many returned to Valley Mountain barefoot from the trials, their feet swollen and bloodied. Recalled one Tennessee veteran: "I was never so completely exhausted and worn out in my whole life."

"*In the subsequent campaigns of the Army of Northern Virginia the troops were subjected to great privations and to many severe trials—in hunger often; their nakedness scarcely concealed; strength at times almost exhausted—but never did I experience the same heart-sinking emotions as when contemplating the wan faces*

Sam Watkins

General Lee on Traveller

and the emaciated forms of those hungry, sickly, shivering men of the army at Valley Mountain!"—**Walter Taylor, aide to General Lee**

By late September, much of the army retreated south. Defeat in these mountains damaged General Lee's reputation, yet he learned valuable lessons in leadership. Lee's legendary rapport with Confederate troops began here.

"One evening, General Robert E. Lee came to our camp...He had a calm and collected air about him, his voice was kind and tender, and his eye was as gentle as a dove's. His whole make-up of form and person, looks and manner had a kind of gentle and soothing magnetism."—**Sam Watkins, 1st Tennessee Infantry**

Fun Fact: General Robert E. Lee grew his trademark beard while campaigning in these mountains. From nearby Greenbrier County, Lee acquired Traveller, the famous warhorse that carried him safely through the rest of the Civil War.

Directions: You might end the day by visiting Snowshoe Mountain resort, a four-season destination that offers adventure-filled vacations. Shopping, restaurants, and lodging can be found nearby.

END OF DAY 2

DAY 3

CHEAT MOUNTAIN PASS, CHEAT SUMMIT FORT, BATTLE OF GREENBRIER RIVER, BATTLE AT CAMP ALLEGHENY

Preview: This tour highlights drama along the Staunton-Parkersburg Turnpike from July-December, 1861. In mid-July, Federal troops seize the pass over Cheat Mountain and build an "impregnable" fortress upon the crest. Meanwhile, Confederates led by General Robert E. Lee move against them. Lee's expedition in this rugged wilderness goes badly, and his first command ends in stunning defeat.

By October, Federal troops take the offensive. The army of Union General Joseph Reynolds marches against Confederates blocking the Staunton-Parkersburg Turnpike at Camp Bartow, on the headwaters of Greenbrier River. Amid dazzling autumn colors, a fierce artillery duel takes place at the Battle of Greenbrier River. Reynolds fails to breach the Confederate defenses and retreats to Cheat Mountain—a rare success for Southern arms in Western Virginia.

In mid-December, Federal troops again push east on the turnpike, with their eye on the Shenandoah Valley. This time, newly minted General Robert Milroy assaults the Confederates, now fortified atop Allegheny Mountain.

The battle at Camp Allegheny is one of the fiercest of 1861. Although badly outnumbered, the Confederates claim victory, sending General Milroy's bloodied bluecoats back to Cheat Mountain. The clash dooms both armies to a severe winter in the Alleghenies. The drama of human triumph and sacrifice, a wilderness setting and stunning mountain scenery await you on **DAY 3**.

A Note on Weather: Portions of Day 3 will be spent at altitudes of more than 4,000 feet above sea level. Even in summer, the temperature at these heights can be markedly cooler—and the weather more severe—than in the valleys below. Please dress accordingly!

Directions: Your Day 3 tour begins at the intersection of US 219/250 in Huttonsville. If approaching from the Snowshoe Mountain area, or points south (US 219), turn right on US 250, and continue south. If approaching from Elkins or points north, proceed through the intersection and continue south on US 250.

You will immediately cross a bridge over the Tygart Valley River. Proceed south on US 250, passing large agricultural fields and the

DAY 3
Cheat Mountain Pass • Cheat Summit Fort • Greenbrier River • Camp Allegheny

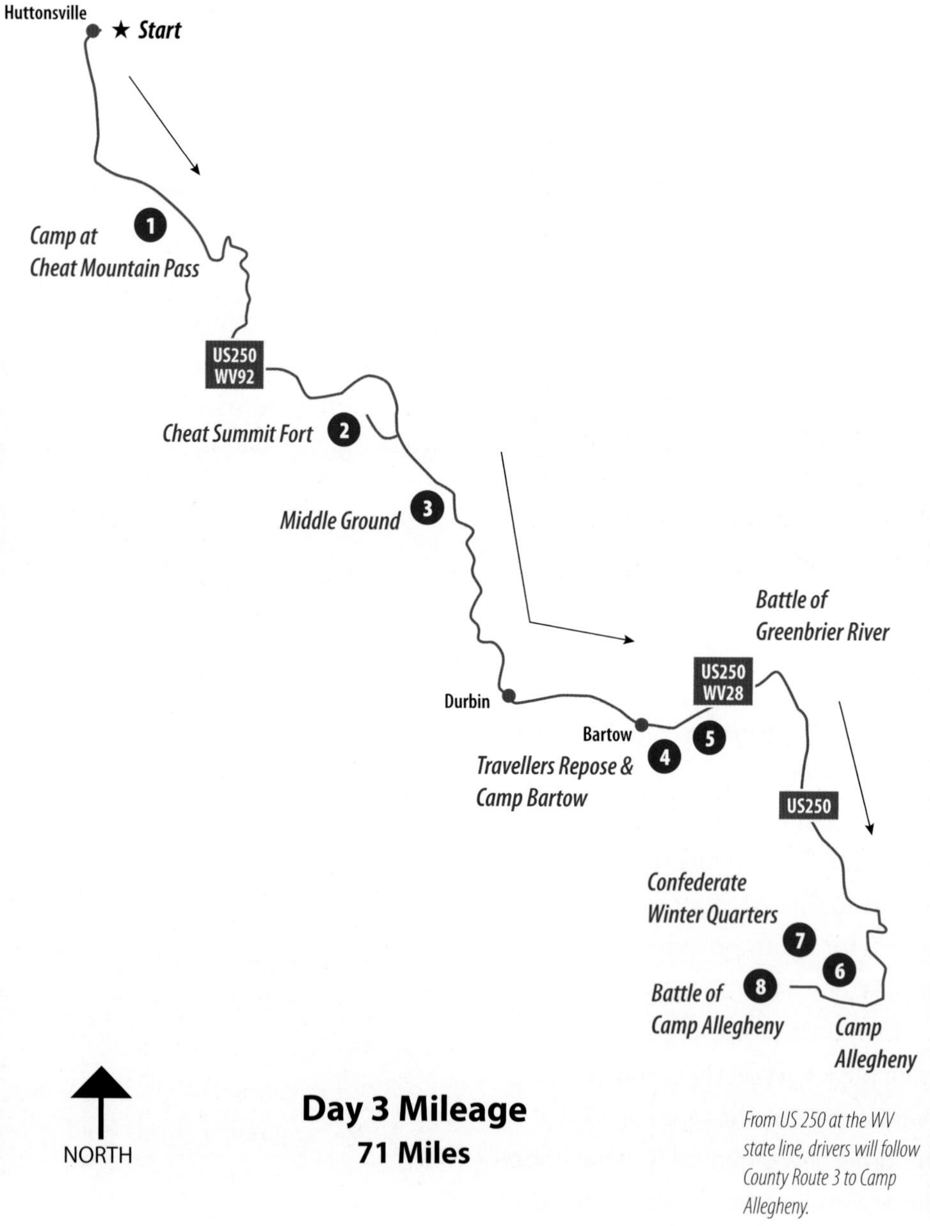

Huttonsville ● ★ *Start*

① *Camp at Cheat Mountain Pass*

US250 WV92

Cheat Summit Fort ②

Middle Ground ③

Battle of Greenbrier River

US250 WV28

Durbin ●

Bartow ●

④ ⑤

Travellers Repose & Camp Bartow

US250

Confederate Winter Quarters

⑦

⑥

Battle of Camp Allegheny ⑧

Camp Allegheny

↑ NORTH

Day 3 Mileage
71 Miles

From US 250 at the WV state line, drivers will follow County Route 3 to Camp Allegheny.

Union headquarters at
Cheat Mountain Pass

Stop 1—Camp at Cheat Mountain Pass

Huttonsville Correctional Center. In the fields to your right was an important Federal camp at "Cheat Mountain Pass" (2.7 miles). You may pull off on the right side of the road.

This site was occupied by Federal troops after the Battle of Rich Mountain. By July 16, 1861, Federals seized the Staunton-Parkersburg Turnpike on top of Cheat Mountain, nine miles southeast. General Joseph J. Reynolds, commanding the First Brigade, "Army of Occupation," made headquarters under canvas here and established a camp to supply and reinforce the fortifications under construction on Cheat Mountain and in the Tygart Valley at Elkwater.

"Here we are cooped up in a narrow gorge, three companies of the 14th Indiana, the whole of the 15th Indiana, and 3d Ohio, a battery of six pounders from Coldwater, Mich, and a company of German cavalry from Cincinnati." — "Prock," 14th Indiana Infantry

A giant blackberry patch behind Reynolds' camp made it notable for tasty pies and cobbler—and for rattlesnakes. The snakes seemed to be fond of crawling inside the soldiers' tents and bedding. Watch your step!

Union regiments pitched rows of white tents in the fields to your right (south) along the aptly named Riffle Creek. On the crest of the steep wooded ridge to your left waved a large American flag, visible for miles in every direction. At this camp, Lieutenant Colonel John Beatty recalled the first of many night alarms:

"The sound of a musket is just heard on the picket post, three quarters of a mile away, and the shot is being repeated by our line of sentinels. The whole camp has been in an uproar. Many men, half asleep, rushed from their tents and fired off their guns in the company grounds....The tents were struck, Loomis' Michigan Battery manned, and we awaited the attack, but none was made. It was a false alarm. Some sentinel probably halted a stump and fired, thus rousing a thousand men from their warm beds."

One of the most unusual meetings of the war took place here. On July 18, more than four hundred unarmed Confederates—paroled prisoners from Rich Mountain—bivouacked in a nearby meadow. That night, some of the Federals walked over to visit their camp. Soldiers of the two armies gathered around campfires and discussed the war, its causes, and why they were compelled to fight.

Directions: Continue traveling south on US 250. A sweeping right turn will begin your ascent of Cheat Mountain (5.3 miles). The road closely follows the old Staunton-Parkersburg Turnpike grade, twisting and turning for more than three miles as you climb toward the summit. Drive carefully.

In 1861, huge rocks and overhanging trees shaded this roadway, adding mystery to the ascent. The difficult march up Cheat Mountain was "enough to dampen the military ardor of almost any one;" wrote a soldier, "but, like the man who carried the calf until it grew to be an ox, we have got accustomed to it."

*"As we climbed the Cheat, the views were the grandest I ever looked upon....The road is crooked beyond description, but very solid and smooth."—**Lieutenant Colonel John Beatty, 3rd Ohio Infantry***

Proceed to the top of Cheat Mountain. Upon reaching the crest, the road straightens out considerably (8.6 miles). Note the change in vegetation; tall red spruce trees and rhododendron are trademarks of the mountaintop.

Cheat Mountain is a remarkable place. In 1861, this region was an authentic wilderness. Much of the mountain remains so today. Few ventured here until the Staunton-Parkersburg Turnpike was completed in the 1840s. Famed Harvard botanist Asa Gray used the new road to explore in this vicinity and discovered plants unknown to science. A new species of minnow and salamander have also been found.

Another curiosity is the Shavers Fork of Cheat River, a stream of considerable size that glides along the *top* of the mountain. One amazed soldier wrote that it "was undoubtedly placed there by a mistake of nature."

☞ **Fun Fact:** The weather on Cheat Mountain is legendary. Rain and snow fall here in prodigious quantities. During the winter of 1855, the Trotter brothers had a contract to carry mail over this mountain from Staunton, Virginia. At one point, a severe snowstorm brought delivery to a halt. When complaints reached the Postmaster General in Washington, the brothers wrote:

"Sir, If you knock the gable end out of Hell and back it up against Cheat Mountain and rain fire and brimstone on it for forty days and forty nights, it won't melt the snow enough to get your d___ mail through on time."

Directions: Continue south on US 250. The old turnpike grade departs from the modern road along this stretch (it winds up the mountainside on your right to the crest of "White Top," out of view). Shavers Fork will soon appear on your left.

A sign marks the gravel road to Cheat Summit Fort on the right (12.5 miles). Take this right, (just before US 250 crosses a steel bridge over the river) and proceed up the road through the woods. You will soon encounter an intersection (12.9 miles); turn right—placing you back on the original grade of the Staunton-Parkersburg Turnpike. Continue less than a mile to Cheat Summit Fort. Drive carefully; the surface of the old pike can be rocky and uneven. Upon reaching the gap at the mountaintop, turn right into a parking area (13.5 miles).

Welcome to Cheat Summit Fort. This site has been protected and interpreted by the USDA Forest Service, Monongahela National Forest.

Cheat Summit Fort. The Federal soldiers who built this fortress considered it "impregnable."

Stop 2—Cheat Summit Fort

On the afternoon of July 16, 1861, Federal troops seized Cheat Summit Fort. Here the Staunton-Parkersburg Turnpike crossed Cheat Mountain at 4,000 feet in elevation, making it a gateway for invasion and a vital point for defense.

Members of the 14th Indiana Infantry, led by Colonel Nathan Kimball, occupied this rugged ground. "Our tents were pitched on a rocky point," wrote a member of the regiment in his diary that first night, "with a fine forest on every side and a magnificent view of the Alleghenies in front of us, a beautiful romantic, though desolate looking spot."

One of the few habitations on Cheat Mountain was here, a hardscrabble farmstead scratched out of the wilderness by an old mountaineer named Mathias White. Suspected of disloyalty, White and his family were placed under arrest.

Federal scouts began to probe the mountain fastness. Axe-men felled the massive spruces to obtain a field of fire, lobbing the branches to create a barrier of sharp points. The soldiers delighted in their novel assignment. "To one who loves the wildly picturesque in nature," wrote a member of the 14th Indiana, "this region could not fail to awe, to please, to fascinate."

Then began the construction of a unique fortress.

Shavers Fork of Cheat River

Stop 2a

Leave the interpretive sign at the parking area, and walk approximately 100 yards up the trail through the woods (north) to a small clearing. Remains of the original defenses of Cheat Summit Fort can be seen directly to your left (west)—a fern covered trench and embankment. **These remains are fragile—please don't walk on them!**

Federal soldiers created the defenses by stacking spruce logs in crib fashion, then covered them with rocks and dirt from the trench in front to form an embankment. The walls were originally "ten feet high," reckoned one soldier, "eight feet through at the base, narrowing to four feet at the top." These works were linked to fortifications and a stout blockhouse on the opposite (south) side of the turnpike. Protected by artillery and bristling defenses, the men who built this fortress considered it impregnable.

*"That fort surpassed anything of the kind I have since seen, and with our regiment to garrison it, we felt entirely secure."—**David Beem, 14th Indiana Infantry***

Nathan Kimball, colonel of the 14th Indiana Infantry. Seasoned by their trials on Cheat Mountain, his regiment became a cornerstone of the famous "Gibraltar Brigade."

Then the legendary Cheat Mountain weather turned. Cold, driving rains—the same miserable weather that swamped the Confederates on Valley Mountain—wore on the volunteers atop Cheat. "Very wet, cold and disagreeable. Almost as cold as December," scrawled a member of the Fourteenth Indiana upon his diary in mid-August. "We are shivering in an almost winter atmosphere." Believe it or not, on August 13, 1861, it actually *snowed* on this mountaintop!

"While our friends in the states are basking in the sunshine, eating peaches and water-melons, we poor devils are nearly freezing to death upon the top of Cheat Mountain."— **Soldier of the 14th Indiana Infantry**

Making matters worse, few of the men had overcoats. The steady rain and hard service reduced poorly made uniforms to rags. To hide their nakedness, some Federals went on duty wrapped in blankets—like Scotch Highlanders in their kilts. "The name of this mountain certainly could not have been more appropriate," wrote one, "for we have been *Cheated* in various ways...since our arrival." Many fell sick and died here.

Stop 2b

Follow the interpretive trail to the wooden observation platform, just ahead. Carefully walk the steps to the top of the platform. You are now standing in the center of Cheat Summit Fort. The forest boundary surrounding this platform marks the perimeter of the fort. Look closely and you can see the ring of earthworks, just inside the trees.

"Ohio and Indiana now hold this pass....We solicit a visit from General Lee."— **"C.D.," Bracken's Indiana Cavalry**

Robert E. Lee's "Forlorn Hope"

Approximately 3,000 Federal troops were camped here on the night of September 11, 1861 when three Confederate brigades, led by General Robert E. Lee,

Earthworks at Cheat Summit Fort

surrounded the fort. The Confederates had marched through the pathless wilds undetected, yet never launched an assault. At dawn on September 12, Colonel Albert Rust of the 3rd Arkansas Infantry viewed these strong works, but considered it "madness" to attack. Rust's musketry was to be the signal for a coordinated assault. Without it, Lee's plan to overwhelm Cheat Summit Fort was doomed.

The attack devolved into skirmishes in the woods around this fort. Unable to determine numbers in the dense forest, small parties of aggressive Union skirmishers routed entire Confederate brigades. This campaign, Lee's first of the Civil War, ended in defeat (See **DAY 2**, Valley Mountain). The general never filed an official report; privately he dubbed it a "forlorn hope expedition."

Lee left the Cheat Mountain region by late September. Southern critics called him "Granny" Lee, but he had learned valuable lessons of war.

Cheat Summit Fort was a staging area for unsuccessful Federal assaults later in 1861. Log cabins were erected inside these breastworks for winter quarters. The Federals who spent the winter of 1861-62 on this mountaintop endured deep snow and bitter winds.

*"The history of the Rebellion furnishes no instances of greater suffering, excepting in Rebel prisons, than that experienced by the troops on the summit of Cheat Mountain, in the fall and winter of 1861."—**E.C. Culp, 25th Ohio Infantry***

Return to the parking area.

☞ **Fun Fact:** A young woman living with the White family cared for the sick and wounded at Cheat Summit Fort. Soldiers called her the "Maid of the Mist," but she made it clear that bestowal of her heart and hand would only be in exchange for "Linken's Skaalp!"

Directions: Retrace your route along the gravel road to US 250. Turn right on US 250 and proceed southeast to the Randolph-Pocahontas County line. On the left is a scenic overlook (16.4 miles). Carefully pull off on the right side of the road.

Colonel Albert Rust, C.S.A.

Stop 3—Scenic Overlook: The "Middle Ground"

This expansive view overlooks the Greenbrier River headwaters in Pocahontas County, and the rugged mountains along the Allegheny Front leading east to Virginia. During the Civil War, this area was known as the "Middle Ground" due to its location between the armies, and for the divided sentiments of its people. This ground hosted a bitter guerrilla war, led by fearsome "bushwhackers." Families split irreconcilably, civil law collapsed; looting, murder, and mayhem reigned.

During the summer and fall of 1861, Federal outposts were stationed nearby. Sentinels kept a sharp eye out for Confederates and bushwhackers by day, and for prowling mountain lions by night!

"Bushwhackers! a lot of thieving, blood-thirsty moccasin-wearing cut throats (expert woodsmen and mountaineers), crack shots with the rifle, who, too cowardly to fight in the open, would, when outnumbering our pickets three to one, attack them."—**Charles Ross, 13th Indiana Infantry**

Note the rugged terrain. Marching across the Alleghenies was a severe test for Civil War soldiers.

"The road from Staunton is a turnpike cut into the sides and over the tops of the mountains. So tortuous is its course that you may travel for miles without gaining in actual distance more than a few hundred

The *"Middle Ground"*

yards, and sometimes the extremes of our column, stretching out a mile or nearly so in length, would be within a stone's throw of each other."—**Richard T. Davis, 12**[th] **Georgia Infantry**

Directions: From the overlook, carefully return to the roadway and continue southeast on US 250. You will promptly leave the mountain crest and descend into the upper Greenbrier Valley. This road follows the original Staunton Parkersburg Turnpike route for part of the way,

and has many sharp curves. Drive with caution. Continue to the town of Durbin (21.0 miles). Here are shops and a depot for the Durbin & Greenbrier Valley Railroad, a popular excursion train. To learn more, call 1-877-686-7245, or www.mountainrail.com.

Proceed to Bartow. These towns were settled after the Civil War. Bartow was named for the Confederate camp erected here in 1861. Continue to the intersection of US 250 and WV 92 (23.9

*Guerrilla riflemen of the Alleghenies—
the dreaded "bushwhackers."*

miles). Go straight through the intersection on WV 92, and take an immediate left onto the one-lane paved road, County Route 3. Pull off here, across from the prominent home and interpretive signs nearby. Welcome to Camp Bartow.

Stop 4—Camp Bartow: Confederate Defense of the Upper Greenbrier Valley

Stop 4a—Travellers Repose

The paved, one-lane road (County Route 3) is the original grade of the Staunton-Parkersburg Turnpike. On the pike, immediately to your left, stands Travellers Repose, once a renowned inn. Tucked in a picturesque valley, this delightful tavern and farmstead offered welcome rest from a bone-jarring stagecoach ride across the mountains.

Innkeeper Andrew Yeager offered comfortable lodging and feasts of mountain trout, mutton, and pancakes smothered in maple syrup—all for the quaint charge of "four pence ha' penny" (six and ¼ cents). Prior to the Civil War, many notables lodged here, including Stonewall Jackson.

The original inn was burned during the Civil War, and rebuilt in 1869. You may wish to view the interpretive signs around Travellers Repose, but watch for traffic. **This is private property, please show respect and do not trespass!**

If you desire, walk along the shoulder of the old Staunton-Parkersburg Turnpike, (County Route 3) past Travellers Repose and the adjoining two-story Peter Yeager house (ca. 1898). Looking northeast, across a narrow ravine, you will view a steep, grassy ridge. Roughly 1/3 of the way below the crest of this ridge is a horizontal line of Civil War earthworks. These

are some of the well-preserved Confederate fortifications at Camp Bartow.

Camp Bartow and Travellers Repose

Stop 4b—Camp Bartow

On August 13, 1861, Confederate forces claimed the strategic ground along East Fork of Greenbrier River at Travellers Repose. Tents soon filled the nearby meadows and defensive works girdled the hills.

Confederate General Henry R. Jackson, commanding the "Monterey line" of the Army of the Northwest, established headquarters here. Southern forces seized this ground to block the Staunton-Parkersburg Turnpike (County Route 3), and a wagon road leading south to Huntersville (WV 92). They now threatened the Federals at Cheat Summit Fort, only twelve miles west.

This encampment was named "Camp Bartow," after Colonel Francis Bartow, a Georgian martyred at the battle of First Manassas. General Henry Jackson, a fellow Georgian, must have influenced the name.

General Jackson's force took part in Robert E. Lee's aborted assault of Cheat

Summit Fort in September 1861. On the morning of October 3, nearly 5,000 Federals under General Joseph Reynolds returned the favor, attacking General Jackson's 1,800 Confederates here in the Battle of Greenbrier River.

"Four hours without intermission cannon replied to cannon, and fairly shook the mountains upon either side, almost deafening us with their echo."—Gus Van Dyke, 14th Indiana Infantry

The Battle of Greenbrier River was largely an artillery duel. Federal cannons, posted near the turnpike, rained shot and shell on this encampment. Confederate soldiers sought cover in their entrenchments, but shells from the bombardment riddled Travellers Repose.

☞ **Fun Fact:** A small kitten, belonging to members of the 23rd Virginia Infantry, pranced on the earthworks during the battle here, oblivious to the storm of death. Whenever a cannonball smacked the dirt nearby, the playful kitten gamboled about it in glee!

Dysentery and Death

General Jackson's command suffered greatly from dysentery, typhoid, and measles while posted here. A large cemetery on a hilltop behind Camp Bartow holds more than 80 unmarked Confederate graves. Most died from disease, rather than bullets. Ambrose Bierce, a member of the 9th Indiana Infantry, later paid tribute to this cemetery in his essay "A Bivouac of the Dead."

"Imagine yourself in the midst of more than five hundred sick and dying comrades, with scorching fevers and parched lips, far from all that is near and dear to them."—Parson Parker, 12th Georgia Infantry

Directions: Return to the intersection of WV 92 and US 250. Turn right at this triangle intersection, and bear right on US 250. Continue for 0.3 miles and pull off on the right at a farm gate (24.2 miles). The large pastures and open hillside on your right (south) were part of Confederate Camp Bartow. Trenches and two earthen artillery emplacements are visible along the adjacent tree line.

Stop 5—Battle of Greenbrier River: "A Touch of Loyal Thunder and Lightning"

The Battle of Greenbrier River was fought in the meadows along both sides of modern US 250. On the morning of October 3, 1861, General Joseph Reynolds led 5,000 Federal soldiers down the turnpike from Cheat Mountain to attack Camp Bartow.

Reynolds placed thirteen pieces

of artillery behind an orchard within the present town of Bartow (no village existed at the time) and opened fire. Six Confederate cannons replied. The storm of shot and shell was both terrifying and grand. More than 1,500 shells were fired into Camp Bartow during this bombardment.

"The balls pass directly over us, bursted over us, and the fragments rattled like rain on our backs. Some fell before us, ploughed great furrows in the earth and blinded us with dirt."—**William Houghton, 14th Indiana Infantry**

"I can't describe my feelings when the battle began. I could but think of you at home so far away & me here in the fight with the balls flying around...thinking that the next moment one might get me."—**Shepherd Pryor, 12th Georgia Infantry**

Union soldiers probed the Confederate left flank, but were repulsed. General Reynolds' troops then advanced toward the Confederate right. Filing across the base of the steep mountainside to the north, they descended into the valley and skirmished with Confederates near the river. As the Federals closed, gunners of Captain Lindsay Shumaker's Virginia artillery on the grassy hill to the south opened fire and stopped them in their tracks. "Distinctly could their officers be heard, with words of mingled command, remonstrance, and entreaty, attempting to rally their battalions into line," wrote General Jackson.

"Of all the infernal inventions of war, it is these shells. They tear men and horses to tatters in an instant, as they fall whizzing among them."—**Correspondent of the Cincinnati Times**

General Reynolds spied Confederate reinforcements streaming down the turnpike toward Camp Bartow. By 1:00 p.m., with his batteries low on ammunition, Reynolds broke off the engagement. Gathering their dead and wounded, the Federals retreated to Cheat Mountain.

Casualties were surprisingly low, despite the fierce bombardment. The Federals had 8 killed and 35 wounded. The Confederates suffered 6 killed, 33 wounded, and 13 missing. General Jackson and his men received a commendation from the Confederate War Department

Shepherd Pryor, 12th Georgia Infantry, C.S.A.

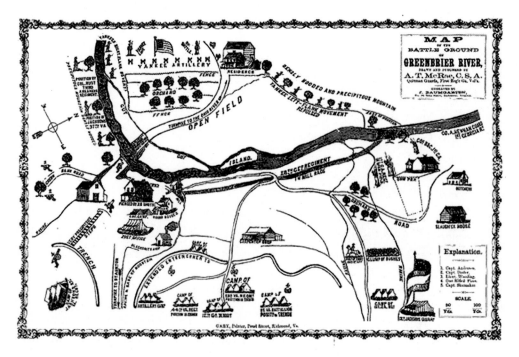

Battle of Greenbrier River,
October 3, 1861

It had his name on it?

James Abbott of the 9th Indiana Infantry died at the Battle of Greenbrier River. "He was lying flat upon his stomach and was killed by being struck in the side by a nearly spent cannon-shot that came rolling in among us," wrote comrade Ambrose Bierce. "It was a solid round-shot, evidently cast in some private foundry, whose proprietor, setting the laws of thrift above those of ballistics, had put his 'imprint' upon it: it bore, in slightly sunken letters, the name 'Abbott.'

for their "brilliant victory." Union General Reynolds was ridiculed for calling his effort merely a "reconnaissance in force."

The Confederates found it difficult to supply troops at Camp Bartow on the deeply rutted Staunton-Parkersburg Turnpike. Therefore, on November 23, they abandoned this camp and retreated nine miles up the muddy pike to the summit of Allegheny Mountain.

Directions: Proceed south on US 250, eventually climbing Allegheny Mountain to the West Virginia-Virginia state line. Road signs will direct you to Camp Allegheny. At the mountain crest (33.1 miles), beside the interpretive signs, turn right on a gravel road leading into the woods. Continue to an intersection (33.7 miles) and turn right—you are again on the original Staunton-Parkersburg Turnpike (County Route 3). Proceed through the woods on County Route 3 to a small gravel parking area and interpretive sign on the left (35.0 miles). Park here and walk to the edge of the tree line facing west. Welcome to the haunting heights of Camp Allegheny.

Stop 6—Camp Allegheny: Overview

In late November 1861, Confederate forces withdrew from Camp Bartow to the mountaintop here at Camp Allegheny. General Henry Jackson departed and Colonel Edward Johnson of the 12th Georgia Infantry took command of the Confederates at this post.

Ed Johnson, a West Point graduate, was a rigid disciplinarian known for his profanity and unquestioned courage. A veteran of the Seminole and Mexican Wars, he won the hearts of his men at the Battle of Greenbrier River. "His manner of fighting was like his speech," thought one Confederate, "no circumvention, no flank movements, no maneuvering for position, no delay—in short, he seemed opposed to taking what might be considered any undue advantage of the enemy."

Colonel Johnson's task was to guard the pass at "Top of Allegheny," less than one mile west. The Staunton-Parkersburg Turnpike winds through that pass at an elevation of more than 4,200 feet.

Face west, and gaze across the expansive grasslands to cleared heights on the near horizon. Those heights, known as Buffalo Ridge, hold most of the Confederate works at Camp Allegheny—the highest Civil War fortification in the east. If you look closely, well-preserved earthworks and artillery emplacements can be seen.

On December 10, 1861, the Confederates at Camp Allegheny received orders to withdraw, in preparation for a winter offensive under Stonewall Jackson. Colonel Ed Johnson called the order a "grave mistake," insisting that Federal troops would follow his retreating Confederates to the Shenandoah Valley.

Three days later, Johnson was vindicated. On December 13, Federal troops under General Robert Milroy struck Camp Allegheny. In a brutal clash fought in bitter cold, Johnson's outnumbered defenders prevailed. The Confederates would remain on this mountaintop all winter.

Camp Allegheny

Ninth Indiana skirmishing in the woods

General Robert H. Milroy U.S.A., the "Gray Eagle."

General Edward "Allegheny" Johnson C.S.A., a profane bulldog in combat

Directions: Return to your vehicle and continue west on County Route 3, following the original grade of the Staunton-Parkersburg Turnpike. For the next mile or so, all the land on your right (north) is Federally owned and part of the Monongahela National Forest. You are welcome to visit this land on foot. All land on your left (south) is privately owned. **Please do not enter any private property without permission!**

On the right, behind a rail fence, you will notice rows of stone mounds in the open field (35.4 miles). Pull off on the right of the turnpike here. If desired, leave your vehicle and walk up to the fence to overlook the field.

Stop 7—Confederate Winter Quarters

In this field, Confederate soldiers built cabins for winter quarters during 1861-62. The stone mounds are the collapsed chimneys from those log shanties. Outlines of the cabins are still visible in the sod.

"Well sis we are into winter quarters at last," penned a member of the 12th Georgia Infantry stationed here on November 28, 1861, "16 men crowded into one little hut—16 ft by 16 ft—one small fireplace to cook, eat, and warm around, and the weather cold, bitter cold, snow all over

87

the ground, and a difficult matter to get wood."

During the battle at Camp Allegheny, Union and Confederate troops fought at close quarters among these structures. One Yankee was shot down upon leaving a cabin with stolen bread. A Rebel who witnessed the act later offered him water and a knapsack for a pillow!

The winter of 1861-62 was bitterly cold in these mountains. Members of the Army of the Northwest suffered terribly at this high altitude post. Frostbite was a constant foe.

*"I have seen ice on the barrels of our guns one forth of an inch thick; I have seen the stoutest men of our regiment wrenching their hands and shedding tears from cold, in short, it's almost a matter of impossibility to describe the sufferings of the soldiers on the Alleghany Mountain."—**Parson Parker, 12th Georgia Infantry***

Cold as the North Pole

Arctic blasts pummeled this exposed encampment. Fierce winds drove smoke down the rude chimneys and sifted snow through clapboard roofs. Long winter days and nights in these shanties were passed with reading, music and preaching—or with smuggled whiskey and gambling. "Men are drunk as usual," wrote a disgusted Confederate in January 1862, "Decent men must endure it—there is no escape."

To avoid the crowded cabins, some Confederates remained in their tents.

"The fireplaces we have constructed do tolerably well while the fire lasts," wrote a member of the 31st Virginia, "but at night we suffer considerably, until the snow blows over us enough to cover us, when we sleep quite well."

*"It is snowing; the wind is blowing a hurricane; it is as cold as the North Pole; and of all the dreary and desolate places on earth, this is entitled to be the palm."—**Anonymous Confederate Soldier***

By spring 1862, the survivors of this encampment were hardy veterans—along with the frostbitten Federals on Cheat Mountain. Those who made the ultimate sacrifice remain buried here in unmarked graves. Please honor them!

*"Sickness is more to be dreaded by far in the army than the bullets. No bravery can achieve anything against it. The soldier may sicken and die, without receiving any attention but from the rough hands of his fellow soldiers."—**James Hall, 31st Virginia Infantry***

Camp Allegheny is amazingly well preserved. Please help protect it!

Directions: Return to your vehicle and continue west on the Staunton-Parkersburg Turnpike (County Route 3). Pull off in a wide spot at the intersection of secondary roads—County Routes 3 and 5

Camp Allegheny

(35.6 miles). The turnpike (Old Pike Road, County Route 3) bears right, descending the mountain to Bartow, nine miles below. The dirt road to the left (Buffalo Mt. Road, County Route 5) climbs Buffalo Ridge, then winds southwest toward the Pocahontas County town of Green Bank. Keep in mind that County Route 5 did not exist during the Civil War.

Stop 8—Battle of Camp Allegheny: December 13, 1861

Union General Robert Milroy, a newly minted brigadier, commanded the Cheat Mountain District of the Department of Western Virginia in December 1861. The impetuous Milroy soon got word from Rebel deserters that the troops at Camp Allegheny were badly "demoralized." Ignoring superstition, he planned to strike them on Friday the 13th.

Milroy patched together a force of 1,900 Federals. On December 12, that force departed Cheat Summit Fort without artillery and marched to Travellers

Repose. There General Milroy divided the Federals into two columns of equal strength, plotting to pounce upon opposite flanks of the enemy camp at dawn. The Confederate defenders of Camp Allegheny numbered only 1,200, led by Colonel Ed Johnson.

Milroy's column started up the icy Staunton-Parkersburg Turnpike at midnight. Less than a mile from this point, they scrambled up the mountainside at first light and struck Confederate pickets on the open ridge to the northeast. The battle was on!

*"Nobody had to waste time hunting up a fight around old Ed Johnson without getting as much as was good for them before night."—**John Robson, 52nd Virginia Infantry***

Colonel Ed Johnson, clad in his nightclothes, grabbed a club of oak root and led the Confederates in a furious countercharge. "The balls flew thick and fast—I really believe that one thousand passed around me, playing all sorts of music," wrote a Confederate soldier. The fighting here was almost hand to hand. "The roar of the musketry was terrible," recalled an Ohio captain, "and the shouts of the men was like the yelling of fiends."

Three times the Confederates were driven to their cabins, but each time they rallied back up the slope, led by the club-waving Colonel Johnson. The Federals found their lines badly thinned, and their ammunition running low. Almost three hours after the attack began, Milroy's column fell back.

*"I fought for more than an hour within a [rod] or two [of] my cabin mates who lay weltering in their blood. It being a very cold frosty morning a fog kept raising from their blood."—**Joseph Snider, 31st Virginia Infantry***

At that point the second Federal column, led by Colonel Gideon Moody of the 9th Indiana Infantry, make their belated appearance on the Confederate left flank, one half mile south. Turning from the wreck of Milroy's column on the right, Colonel Johnson now rushed reinforcements to the Confederate left flank.

Here the Federals found scowling earthworks with acres of slashed timber in front. Charging with a yell, they stalled in the tangle of logs and brush. Confederates rose from the trenches and poured in a murderous fire.

*"It was a horrible sight indeed....I did not see how I could escape their balls. They shot well, and had we not fell down we would undoubtedly have been killed."—**John Chilcott, 9th Indiana Infantry***

Federal troops took shelter in the slashed timber. Some of the bravest crawled to within twenty yards of the trenches. Enemy soldiers taunted and cursed—daring each other to hold up their heads! "In this position we kept up a regular Indian fight for over four hours," recalled a member of the Second (U.S.) Virginia Infantry. "Toward the last the firing became

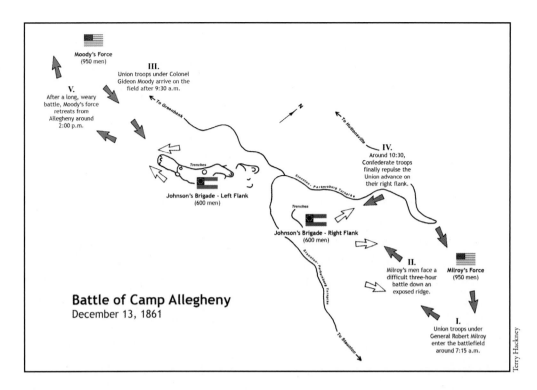

Battle of Camp Allegheny, December 13, 1861

(Map labels:)

Moody's Force (950 men)

III. Union troops under Colonel Gideon Moody arrive on the field after 9:30 a.m.

V. After a long, weary battle, Moody's force retreats from Allegheny around 2:00 p.m.

To Greenbank

N

To Huttonsville

IV. Around 10:30, Confederate troops finally repulse the Union advance on their right flank.

Trenches

Johnson's Brigade - Left Flank (600 men)

Staunton - Parkersburg Turnpike

Trenches

Johnson's Brigade - Right Flank (600 men)

II. Milroy's men face a difficult three-hour battle down an exposed ridge.

Milroy's Force (950 men)

Staunton Parkersburg Turnpike

I. Union troops under General Robert Milroy enter the battlefield around 7:15 a.m.

To Staunton

Battle of Camp Allegheny
December 13, 1861

Terry Hackney

Battle of Camp Allegheny,
December 13, 1861

so accurate, that if an inch of one's person was exposed, he was sure to catch it."

Confederate artillerists splintered the brush with canister to drive out the attackers. By 2:00 p.m., nearly seven hours after General Milroy's first shots, the exhausted Federals withdrew to Cheat Mountain. The Battle of Camp Allegheny was over.

*"Oh, it was heart rending to hear the death shrieking of the dieing, the groans of the wounded and to behold the mangled corpses of the slain."—**Neil Cameron, 25th Ohio Infantry***

A War Department commendation praised the Confederates for victory in "combat as obstinate and as hard fought as any that has occurred during the war." Each side suffered roughly 150 casualties. John Cammack's company of the 31st Virginia Infantry lost eighteen of forty-two men in the battle. "I was a corporal at the time and the command of the company devolved on me," he recalled. "We buried six of our men in one grave, and I commanded the firing party."

It was truly a fraternal fight. Virginia Confederates recognized old neighbors among the Union dead. One Union soldier was said to have drawn a bead on his own brother during that battle!

*"I have seen enough of war. O my God, how forcibly it illustrates the folly and depravity of the human heart."—**James Hall, 31st Virginia Infantry***

Directions for Travel on Foot: You are welcome to explore the grassy hillside behind the fence to the northeast, on property signed and managed by the USDA Forest Service. To the east lie three rows of well-preserved remnants of Confederate cabins. Members of the 31st Virginia Infantry occupied these cabins.

Climbing north along the brow of the cleared ridge above those cabins, you will encounter a Confederate earthwork and some rifle pits. The open ground between the cabins and the ridge crest was the scene of General Milroy's attack on the Confederate right flank. There are fine views from the top of this ridge.

Note: Camp Allegheny is hallowed ground! Please treat it with respect. Do not walk on the earthworks or cabin features. This is Federal land—digging or removal of artifacts is forbidden under penalty of law!

Return to your vehicle. Some of the Camp Allegheny defenses are visible from County Route 5, the narrow dirt road bearing southwest. By walking from the intersection, you can view a well-preserved Confederate artillery emplacement along that roadway on the right of the road, just above a modern stone cabin (0.1 mile). A few additional earthworks can be seen along the left (east) side of the road just beyond.

Remember: This is private property! Do not leave the public roadway for any reason without permission! Be advised: there are few spots nearby suitable for turning a vehicle on County Route 5.

☞ **Fun Fact:** The victory at Camp Allegheny made Colonel Ed Johnson a Southern hero. His "immense war club" went on display at the Virginia State Library, and he received a general's star dated from the day of battle—Friday the 13th! He also earned a lifelong nickname: "Allegheny" Johnson.

Michael Ledden

Camp Allegheny tour

The First Campaign Draws to a Close

While battle raged at Camp Allegheny on December 13, 1861, politicians in Wheeling established the constitution and territorial limits of the proposed new state of "West Virginia." "[W]inter closes in on the Union people of Western Virginia," noted President Abraham Lincoln in his message to Congress that year, "making them masters of their own country."

By April 1862, the Confederates were forced to abandon Camp Allegheny and retreat southeast. Federal troops followed. General Stonewall Jackson's Shenandoah Valley campaign now unfolded.

A Legacy of Memories and Ghosts

The epic scale of America's Civil War doomed the First Campaign to obscurity. Yet the little clashes of 1861 in the Allegheny Mountains of "Western" Virginia have great significance.

The First Campaign was decisive. By its end, the focus of war had shifted in Virginia. General McClellan's invasion rallied wavering westerners to the Lincoln government. Union bayonets allowed loyalists to hammer out a new government in Wheeling to oppose the Confederates in Richmond. On June 20, 1863, that novel act resulted in a new state—West Virginia.

The First Campaign was a proving ground for armies and leaders who shaped American history—George McClellan, Robert E. Lee, and a host of others. Here

93

McClellan won the first Union victories of the war and rocketed to General-in-Chief, while a mud-spattered "Granny" Lee left the mountains in defeat. It might have been the greatest irony of the war.

In the glow of victory, General McClellan's fatal flaws were overlooked. During the next year however, his undue caution and loss of nerve would become legendary. President Lincoln finally sacked him. The talented McClellan organized two great armies, but his rapid rise to high command likely prolonged the Civil War.

General Lee, on the other hand, gained valuable lessons in defeat. He learned to handle raw troops and recalcitrant commanders in Western Virginia. Recalling the bitter disappointments here, Lee resolved to strike boldly in future campaigns. Applying those lessons, he became a legendary military leader.

Soldier life in the rugged Alleghenies chiseled raw recruits into veterans. Many called their First Campaign the "severest" of the war. A surprising number went on to fill the ranks of storied armies North and South.

"The history of that remarkable campaign would show, if truly portrayed, a degree of severity, of hardship, of toil, of exposure and suffering that finds no parallel [and]... would have done honor to our sires in the most trying times of the Revolution." — **Colonel Samuel Fulkerson, 37th Virginia Infantry**

Directions: Return to the intersection of County Routes 3 and 5. To retrace your route to US 250, continue east on County Route 3 for 1.8 miles, turn left at the intersection and proceed another 0.6 miles to US 250 at the West Virginia-Virginia state line.

I hope you've enjoyed the tour!

Michael Ledden

The Severest Campaign

Further reading

Lesser, W. Hunter. *Rebels at the Gate: Lee and McClellan on the Front Line of a Nation Divided.* hlesser@suddenlink. net.

Cohen, Stan. *The Civil War in West Virginia: A Pictorial History.* Charleston, WV: Quarrier Press, 1976.

_____. *A Pictorial Guide to West Virginia's Civil War Sites.* Charleston, WV: Quarrier Press, 1990.

Lowry, Terry and Stan Cohen. *Images of the Civil War in West Virginia.* Charleston, WV: Quarrier Press, 2000.

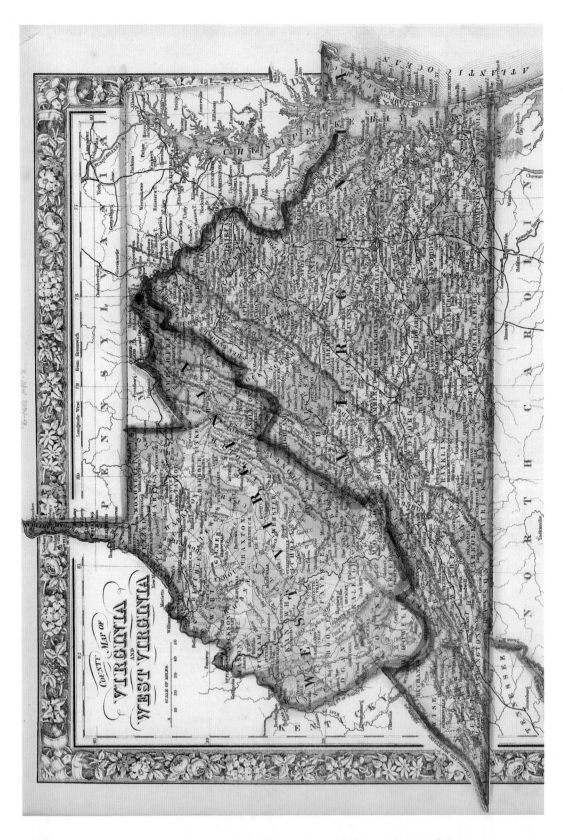

West Virginia statehood (June 20, 1863) was a legacy of the First Campaign